# The Lupus Book

# The Lupus Book

## A Guide for Patients and Their Families

DANIEL J. WALLACE, M.D.

Clinical Chief of Rheumatology
Cedars-Sinai Medical Center
Clinical Professor of Medicine
UCLA School of Medicine
Los Angeles, California

*New York Oxford*
OXFORD UNIVERSITY PRESS
1995

Oxford University Press

Oxford New York
Athens Auckland Bangkok Bombay
Calcutta Cape Town Dar es Salaam Delhi
Florence Hong Kong Istanbul Karachi
Kuala Lampur Madras Madrid Melbourne
Mexico City Nairobi Paris Singapore
Taipei Tokyo Toronto

and associated companies in
Berlin Ibadan



Published by Oxford University Press, Inc.,
198 Madison Avenue, New York, New York 10016



Library of Congress Cataloging-in-Publication Data
Wallace, Daniel J. (Daniel Jeffrey), 1949–
The lupus book / Daniel J. Wallace;
editors, Frances Brock and Janice Wallace.
p. cm. Includes bibliographical references (p. ).
ISBN 0–19–508443–8
1. Systemic lupus erythematosus—Popular works.
RC924.5.L85W35 1995 615.7'7—dc20 94–37621

Dose schedules are being continually revised and new side effects recognized. Oxford University Press and the author make no representation, express or implied, that the drug dosages and recommendations in this book are correct. For these reasons, readers are strongly urged to consult their physicians and the drug company's printed instructions before taking any drugs.

3 5 7 9 8 6 4

Printed in the United States of America
on acid-free paper

# Foreword by Henrietta Aladjem

Cofounder, Lupus Foundation of America
Editor, *Lupus News*
Watertown, Massachusetts

As someone once said, the story of lupus is one that we should know more about. For patients who want and need information about their disease, who want to take charge of their lives in the face of illness, and who want the ability to carry on an intelligent discussion of treatment with their physicians, reading *The Lupus Book* is an important step.

When I was first diagnosed with lupus in 1953, I scanned a few medical libraries for facts about the disease. It should have been relatively easy for me to turn up some information, since I had worked at Widener Library in Cambridge, Massachusetts for several years and had an understanding of how such things were categorized. Yet the search originally yielded only a single book, published by the Finsteen Institute in Denmark, and this tiny publication dealt only with the worldwide prevalence of lupus and tuberculosis. I conveyed my dismay and chagrin at this lack of information to several reference libraries. It was apparent that few physicians were interested in writing about this disease, which had rather suddenly become my disease.

Through all these years, I never came across a book about lupus written in language simple enough for patients to understand. As a matter of fact, medical jargon is becoming so complicated that even doctors are finding it hard to communicate with one another.

The lupus patient can easily become bewildered, suffering not only from the relentless attack of the disease, but also from the fear of death and dying and the lack of understanding about what this disease can and will do to a human life. Now, for the first time, *The Lupus Book* will describe to the patient in lay language the latest medical findings about this disease and the treatments designed to ameliorate it. Patients, their families, and their friends will benefit from *The Lupus Book,* and so will nurses, social workers, pharmacists, dentists, and mental health workers or anyone else who wants to know more about it.

Education of the medical community at large about lupus is of critical importance. We need to ease the burden of the rheumatologist and immunologist, who often do not have sufficient time to deal with the multisystem problems of their lupus patients. Such practitioners will have much more success dealing with informed patients who have confidence in their potential to help themselves and willingly comply with prescribed medications and treatments.

Some health organizations, such as the Arthritis Foundation, The American Lupus Society, and the Lupus Foundation of America have made attempts to educate the lupus patient. Each year, they mail thousands of easy-to-understand, well-researched educational pamphlets and medical papers to patients all over the world, in many different languages. However, *The Lupus Book,* with all the information one might need close at hand, will prove a blessing in that it summarizes and makes readable all the pertinent information in a single source.

Today, there are hundreds, perhaps thousands of papers on immunology, autoimmunity, and lupus, and there are quite a few books on the subject. As the clouds of darkness break and a highly promising blue sky illuminates the scientific horizon, patients everywhere are beginning to feel more hopeful. They are hoping for new cures, medications with less side effects, a better quality of life, and perhaps a cure in our lifetime, so that we can call an end to so much suffering and so many unnecessary deaths.

Had there been a book like *The Lupus Book* when I had active lupus (fortunately, I have been in complete remission for 25 years), much suffering would have been spared, not only for myself but for my spouse and children as well. *The Lupus Book* is an important addition to the patient-oriented literature on lupus and takes its place alongside its sister textbook *Dubois' Lupus Erythematosus,* by Drs. Wallace and Hahn.

# Preface

I'm amazed at the number of lupus patients referred to me who have received only a cursory explanation of their disease and a brief discussion of its management. They have no idea what to expect and therefore usually have many questions, some of which I cannot yet answer. I have written this book for them and their physicians. Rheumatology textbooks and lupus monographs are available at medical libraries, but the information in them is not presented in a way that patients and their families can easily understand. The Arthritis Foundation and lupus support groups (e.g., The American Lupus Society and the Lupus Foundation of America) publish excellent pamphlets on various aspects of the disorder, but these are often superficial; they do not explain in detail the disease's mechanisms or put therapies in their proper context. Several books have appeared for the lay audience, but with one notable exception, they are either outdated, describe personal struggles, or concern themselves with promoting coping strategies. (The exception are the books by Henrietta Aladjem; the Appendix lists her publications and her foreword to this book precedes this preface.) Less comprehensive monographs by physicians have appeared.

As a physician who specializes in rheumatology and has a special interest in lupus, I have tried to anticipate your questions with the most up-to-date information we now have on causes, prevention, cure, exercise, diet, and many other important topics. This book is a distillation of my experience in treating over a thousand lupus patients.

*The Lupus Book* is in many ways a lay companion to *Dubois' Lupus Erythematosus,* which I coauthored with Bevra Hahn. This 955-page textbook contains well over eight thousand references and is considered one of the most comprehensive works on the subject. I have duplicated the organization and structure of this textbook here in an attempt to "translate" it for my patients. In doing so, I have also kept in mind the many allied health professionals (physical therapists, nurses, occupational therapists, social workers, psychologists, and others) who may be involved in the detection and care of lupus patients.

I hope all of you find this work informative and enjoyable to read. If you are reading this book, you may have been diagnosed with lupus or suspect that you have it. It is my hope that this book will help you work with your physician. In some instances, to flesh out the details, I have used composite cases based on real people I treat. Of course—as I learned early in my practice—no two patients have exactly the same experience with this disease. But some of my patients' personal stories may ring true and help you cope with your own symptoms.

My book is not intended to be a substitute for advice given by your family physician or the specialist you have been referred to. These doctors know your medical history and related problems far better than I ever will and can provide you with a perspective that it is not possible for me to impart.

*The Lupus Book* is not meant to be read from front to back. It is intended as a resource for patients and caregivers who are interested in how various aspects of the disease are approached. In particular, Chapters 5 to 9 may be very technical. The reader should not get discouraged; understanding immunology is a daunting task even for physicians.

I wish to thank Lea & Febiger, my textbook publisher, for allowing me to use materials from *Dubois' Lupus Erythematosus* for this publication. Special thanks are owed to Ruth Wreshner, my editor Frances Brock, Allan Metzger, M.D., Nancy Horn, my medical artist Terri Hoffman, Joan Bossert and the people at Oxford University Press, as well as my wife (and editor) Janice, and our three children, Naomi, Phillip, and Sarah.

Additional thanks are owed to my long suffering secretary, Amanda Trujillo, the Lupus Foundation of America (especially John Huber, Kerryn Coffman, and Judy Madwin), The American Lupus Society (especially Leslie Epstein), and Drs. R. H. Phillips and Jim Maguire for their inspiring writings on lupus and rheumatoid arthritis.

*Los Angeles* D. J. W.
*April 1995*

# Contents

# Part I
# INTRODUCTION AND DEFINITIONS

Where should we start? The most logical place is with a definition of lupus. We look at how it is classified as a disease and place it in its proper historical perspective. This is followed by an overview of how lupus is distributed in the population—in other words, who gets the disease, which parts of the world have the highest prevalence of lupus, how many people have lupus in the United States, at what age, and which sex is most affected.

# 1

# *Why Write a Book on Lupus?*

The first time someone hears the words ''lupus erythematosus,'' he or she usually says *''What?''* When I first started my practice, patients identified the term with Peter Lupus, one of the characters on *Mission Impossible,* a popular television series in the late 1960s. Sometimes it looks as though finding a cure for lupus is an impossible mission, but there is much we *do* know, and the aim of this book is to share that knowledge.

Lupus is the common name for the disorder known technically as lupus erythematosus. This formal name includes systemic lupus erythematosus—where *systemic* means affecting the entire body or internal system—or SLE for short. Although underrecognized, lupus is an extremely important disease for many reasons:

- *In the United States, nearly one million people suffer from lupus.* It is more common than better-known disorders such as leukemia, multiple sclerosis, cystic fibrosis, and muscular dystrophy *combined.* Those who develop SLE do so in the prime of life. And 90 percent of these sufferers are women, 90 percent of whom are in their childbearing years. Moreover, the effects of the disease disrupt family life and account for hundreds of millions of dollars in lost productivity.
- *Understanding the immunology of lupus will help us better understand AIDS, infections in general, allergies, and cancer.* Medical students are often told, ''Know lupus and you know medicine.'' This is because SLE can affect every part of the body. The basic pathology of lupus, or the factors that cause the disease, get to the core of how the human immune system functions. Nearly every major advance in understanding lupus immunology has had a spillover effect—it has helped not only SLE patients but also those with immune-related disorders such as allergies, cancer, AIDS, and other infectious processes.
- *Lupus can be a very difficult disease to diagnose.* Many lupus patients look perfectly healthy, but surveys have shown that newly diagnosed patients have had symptoms or signs for an average of 3 years. A young woman who complains of fatigue, achiness, stiffness, and low-grade fevers or swollen glands is often told she is experiencing stress, has picked up a virus that is going around, or—worse—that she is exaggerating her symptoms. By the

time she is diagnosed with SLE, permanent damage to vital organs such as the lungs or kidneys may have occurred. (Serious lupus is usually easy to diagnose.) This book attempts to increase public awareness of the disease, which could lead to earlier diagnosis.

- *The diseases of females are understudied by organized medicine.* For years, many medical protocols have tended to limit funded studies to males. A survey done in the late 1980s showed that 70 to 80 percent of all research participants in treatment protocols being conducted in the United States were men. (Some of this bias can be explained by the preferential funding given to Veterans Administration hospitals.) But diseases that primarily affect females are funded to a lesser extent than other less common disorders, such as leukemia or muscular dystrophy. If the population of patients suffering from lupus were 90 percent male, I daresay that the medical community would be more responsive. Research on lupus is also relatively underfunded compared to studies of other life-threatening diseases.
- *It is my opinion that there is a shortage of doctors capable of diagnosing and treating SLE,* a disease studied and managed by rheumatologists. Rheumatology is one of the recognized subspecialties of internal medicine, along with cardiology, gastroenterology, and pulmonary medicine, but it was certified only in 1972. It is therefore a relative newcomer—a field in which only 3000 of the 550,000 physicians in the United States are certified to practice.
- *Many patients who are told they have SLE do not.* Some ten million Americans have a positive lupus blood screen (called antinuclear antibody, or ANA) but only about one million of these actually have SLE. Since normal patients and healthy relatives of those with autoimmune disease can have positive tests for lupus, some physicians take the test results at face value and inform their patients (especially young women) that they do indeed have the disease or may succumb to it in future. Such patients may suffer ill effects, especially if unnecessary treatments are prescribed. Also, many disorders mimic SLE. A positive blood test for lupus may be found during a viral illness, and unsuspecting physicians may draw the wrong conclusions. Disorders closely related to SLE, such as scleroderma or polymyositis (see Glossary for definitions of technical terms), may exhibit similar test results but are treated quite differently. In approaching this difficult diagnosis, a complex diagnostic workup is often necessary, and few physicians are equipped to interpret the necessary battery of tests. In these instances, most physicians will consult a board-certified rheumatologist or recommend that their patients visit such a specialist.

Now let's get started—and we'll begin by discussing what lupus really is.

# 2
# *What Is Lupus?*

In simple terms, lupus erythematosus develops when the body becomes allergic to itself. Immunologically speaking, it is the opposite of what takes place in cancer or AIDS. In lupus, the body overreacts to an unknown stimulus and makes too many antibodies, or proteins directed against body tissue. Thus, lupus is called an *autoimmune disease* (*auto-* meaning *self*).

## IS THERE AN "OFFICIAL" DEFINITION OF LUPUS?

The American College of Rheumatology (ACR), a professional association to which nearly all rheumatologists in the United States belong, devised criteria for defining the disease in 1971. These criteria were revised in 1982 and are shown in Table 1. The presence of 4 of the 11 criteria confirms the diagnosis. These criteria apply only to SLE and not to drug-induced or discoid (cutaneous) lupus. (These various forms of lupus are discussed under the next heading.)

The first four criteria concern the skin: sun sensitivity, mouth sores, butterfly rashes, and discoid (resembling a disk) lesions.

The second four criteria are associated with specific organ areas: the lining of the heart or lung, the kidneys, the central nervous system, and the joints.

The remaining three criteria specify relevant laboratory abnormalities: altered blood counts (low red blood cells, white blood cells, or platelets), positive ANA (antinuclear antibody) testing, and other blood antibody abnormalities of the disease. The ANA test is used as the primary diagnostic tool to determine whether a person has lupus, but there are limits to its reliability, which we discuss in Chapter 6.

Many other manifestations of SLE are not included in the ACR criteria. They are excluded because they are not *statistically* important in differentiating SLE from other rheumatic diseases. For example, a condition known as Raynaud's phenomenon (when one's fingers turn white and then blue in cold weather) is present in one-third of lupus patients. But it is not included in the criteria, since 95 percent of those suffering from scleroderma also have Raynaud's. In other words, it is not specific to SLE and therefore does not provide enough proof to classify someone as having SLE. These particular manifestations of SLE will be covered in detail in later chapters.

**Table 1.** *ACR (1982) Revised Criteria for the Classification of Systemic Lupus Erythematosus*

A person is said to have SLE if four of the eleven following criteria are present at any time:

*Skin criteria*

1. Butterfly rash (lupus rash over the cheeks and nose)
2. Discoid rash (a thick, disklike rash that scars, usually on sun-exposed areas)
3. Sun sensitivity (rash after being exposed to ultraviolet A and B light)
4. Oral ulcerations (recurrent sores in the mouth or nose)

*Systemic criteria*

5. Arthritis (inflammation of two peripheral joints with tenderness, swelling, or fluid)
6. Serositis (inflammation of the lining of the lung—also called the pleura—or the heart—also called the pericardium)
7. Kidney disorder (protein in urine samples or abnormal sediment in urine seen under the microscope)
8. Neurologic disorder (seizures or psychosis with no other explanation)

*Laboratory criteria*

9. Blood abnormalities (hemolytic anemia, low white blood cell counts, low platelet counts)
10. Immunologic disorder (blood testing indicating either a positive LE cell preparation, anti-DNA, false-positive syphilis test or a positive anti-Sm)
11. Positive ANA blood test

## WHAT TYPES OF LUPUS ARE THERE?

Sometimes the autoimmune reaction of lupus can be limited just to the skin and may result in a *negative* ANA blood test. This condition is called *cutaneous* or *discoid lupus erythematosus* (*DLE*). Though this is not an entirely accurate term (see Chapter 12), it helps distinguish these patients from those suffering with systemic lupus. About 10 percent of lupus patients exhibit this condition. When internal features are also present and fulfill ACR criteria (Table 1), we describe the condition as *systemic lupus erythematosus* (*SLE*).

SLE patients who have symptoms of achiness, fatigue, pain on taking a deep breath, fevers, swollen glands, and signs of swollen joints or rashes but whose internal organs are not involved (for example, the heart, lung, kidney, or liver) are said to have *non-organ-threatening disease*. Statistics vary, but on the basis of my own clinical experience, I estimate that about 35 percent of lupus patients fall into this category. Patients with non-organ-threatening disease have a normal life expectancy, and it is uncommon for them to develop disease in the major organs after the first 5 years of having the disease.

On the other hand, involvement of the heart, lungs, kidneys, or the presence of liver or serious blood abnormalities indicates that an *organ-threatening disease* is at work. This may become life-threatening if the patient is not treated with

**Table 2.** *Types of Lupus Erythematosus*

| |
|---|
| Cutaneous (discoid) lupus erythematosus (10%) |
| Systemic lupus erythematosus (70%) |
| Non-organ-threatening disease (35%) |
| Organ-threatening disease (35%) |
| Drug-induced lupus erythematosus (10%) |
| Crossover or overlap syndrome and/or MCTD (10%) |

corticosteroids or other interventions. Another 35 percent of all lupus patients fall into this category.

Approximately 10 percent of patients with lupus develop the disease for the first time from a prescription drug and have what is called *drug-induced lupus erythematosus*. Only about half of this group fulfills the ACR criteria for SLE. The drug-induced form is usually less severe than SLE and will disappear after the patient stops taking the particular drug. Occasionally, however, short courses of lupus medication are required for these patients.

Finally, perhaps 5 to 10 percent of the individuals who fulfill the ACR criteria for SLE may also fulfill the ACR criteria for another autoimmune disorder such as scleroderma (tight skin with arthritis), dermato-/polymyositis (inflammation of the muscles), or rheumatoid arthritis (a potentially deforming joint inflammation). These patients are said to have *mixed connective tissue disease* (*MCTD*) if they possess a particular autoantibody (anti-RNP). If they do not, the patients are said to have a *crossover* or *overlap syndrome*. This classification system is summarized in Table 2.

## WHAT'S IN STORE FOR THE READER

Don't be overwhelmed by all these facts and figures. This chapter has simply provided you with an overview of the book, and all the points mentioned will be discussed again in more detail in later chapters.

We close this first part with a brief historical background and an overview about who gets lupus (Chapters 3 and 4). In Part II, the heart of the book, we look at the immune system and how it relates to SLE (Chapters 5 to 9). We discuss the manifestation of the disease in different areas of the body, such as the joints, the gastrointestinal system, the kidneys, and other organs (Chapters 12 to 20) and talk about the role of blood testing (Chapter 11). I explain the necessary clinical and diagnostic studies (x-rays, scans, etc.) that are used in assessing lupus (Chapters 12 to 20), as well as problems unique to specific circumstances, such as pregnancy, infection, and lupus in children and the elderly (Chapters 22, 23, 29, and 30). Next, we take up the treatment of lupus—the physical measures we can take to combat the disease, the various medications, and the emotional support you will need from your family and physician (Chapters 24 to 28). Finally, future directions and advances soon to take place are detailed in Chapters 31 and 32.

# 3
# *The History of Lupus*

''Lupus'' is the Latin word for ''wolf,'' and it is common medical lore that the ''butterfly rash'' seen on the cheeks of many lupus patients is so similar to the facial markings of a wolf that our ancestors chose the name for this reason. The technical name for the disease we know of as lupus—lupus erythematosus—was first applied to a skin disorder by a Frenchman, Pierre Cazenave, in 1851, though descriptive articles detailing the condition date back to Hippocrates in ancient Greece.

Accurate treatises on the skin disorders associated with lupus were published in the mid-1800s by the great Viennese physicians Ferdinand von Hebra and his son-in-law Moriz Kaposi (for whom Kaposi's sarcoma is named). The first suggestions that the disease could be internal (more than skin deep and affecting the organs of the body) appeared in these writings. However, it was Sir William Osler (the founder of our first real medical internship and residency programs in the 1890s at Johns Hopkins) who wrote the earliest complete treatises on lupus erythematosus between 1895 and 1903. In addition to describing such symptoms as fevers and aching, he clearly showed that the central nervous, musculoskeletal, pulmonary, and cardiac systems could be part of the disease.

The golden age of pathology in the 1920s and 1930s led to the first detailed pathologic descriptions of lupus and showed how it affected kidney, heart, and lung tissues. Early discussions of abnormal blood findings such as anemia (low red blood cell count or low hemoglobin) and low platelet count (cells that clot blood) appeared during this time. We had to wait until 1941 for the next breakthrough, which took place at Mount Sinai Hospital in New York City. There, Dr. Paul Klemperer and his colleagues coined the term ''collagen disease'' on the basis of their clinical research. Although this term is a misnomer (collagen tissues are not necessarily involved in lupus), the evolution of this line of thinking led to our contemporary classification of lupus as an ''autoimmune disorder,'' based on the presence of ANA and other autoantibodies.

The first arthritis unit with a special interest in lupus was started by Marian Ropes at the Massachusetts General Hospital in Boston in 1932. In those days, no blood test to diagnose lupus was available. In fact, until 1948, there were no effective treatments for lupus except for local skin salves or aspirin. Dr. Ropes observed that half of her patients got better and half of them died during the first 2 years of treatment. Indirectly, she was classifying her patients into ''organ-threatening'' and ''non-organ-threatening'' categories, but in many cases she had no way short of a tissue biopsy to determine which subset a patient belonged to.

In 1946, a Mayo Clinic pathologist named Malcolm Hargraves performed a bone marrow examination on a patient and absentmindedly kept a tube from the procedure in his pocket for several days. In a bone marrow examination, the physician removes a tissue sample from bone (usually from the sternum or pelvis, where blood components are made). After finally retrieving the tube, Hargraves observed a unique cell on his microscope slides, which became known as the LE cell. Published in 1948, his description of the LE, or lupus erythematosus, cell was one of the landmark developments in the history of rheumatology. This cell was representative of the systemic inflammatory process; its identification allowed doctors for the first time to diagnose the disease faster and more reliably. Dr. Hargraves and others were quick to show how LE cells could be looked for in peripheral blood samples and found that 70 to 80 percent of patients with active SLE possessed these cells. At long last, patients with the disease could be readily identified. Researchers were on a roll: in the following year, 1949, another landmark event took place. Dr. Phillip Hench, another Mayo Clinic physician and the only rheumatologist ever to win the Nobel Prize in Medicine, demonstrated that a newly discovered hormone known as cortisone could treat rheumatoid arthritis. This hormone was administered to SLE patients throughout the country, and immediately dramatic lifesaving dramas took place.

The final chapter of our story evolved during the 1950s, when the concept of autoimmune disease was formalized and the LE cell was shown to be part of an antinuclear antibody (or ANA) reaction. This led to the development of other tests for autoantibodies, which enabled researchers to characterize the disease in a more detailed and definitive manner. My mentor, Dr. Edmund Dubois, amassed an incredible 1000 patients with lupus and was among the first researchers to explore the natural course of the disease and advise how best to treat it. Also during this time, cancer chemotherapy agents such as nitrogen mustard were shown to be effective in the management of serious organ-threatening complications of SLE when used together with corticosteroids.

With this historical context in mind, we now turn our attention to the 1990s and a discussion of who gets lupus and why.

# 4
# *Who Gets Lupus?*

How many lupus patients are there in the United States? It is not as easy to answer this question as it might seem. In 1987, the National Arthritis Data Workshop estimated that there were as few as 128,000 Americans with SLE. These numbers, however, do not include those patients who have discoid lupus or drug-induced lupus. On the other hand, the Lupus Foundation of America, the American Lupus Society, and the Arthritis Foundation have suggested that between 500,000 and 1 million Americans have one of the four forms of lupus (see Table 2).

There are several reasons for these discrepancies. First of all, some epidemiologic (epidemiology is the study of relationships among various factors that determine who gets diseases) surveys assumed that all lupus patients were hospitalized over a 7- to 10-year period and thus gathered their data from hospital discharge diagnoses only. But other groups have shown that less than 50 percent of lupus patients are hospitalized over a 10-year follow-up observation period. Second, data banks from prepaid health plans such as Kaiser-Permanente can track outpatient diagnoses but generally include only patients who are insurable and working. Moreover, many physicians do not list lupus as a diagnosis on an insurance form, since it might result in the policy being canceled or the illness being disclosed to fellow employees. Third, surveys conducted by the Mayo Clinic include a greater than 95 percent Caucasian population, which does not reflect the true racial makeup of the United States or of the disease. Finally, drug-induced lupus lasts only a few weeks in most patients and is infrequently recorded. In addition, discoid lupus patients often see only dermatologists and are rarely hospitalized, which makes it difficult for a rheumatology registry to estimate its prevalence. And again, lupus is often not properly diagnosed.

In spite of these misgivings about underestimates, published surveys in the United States of mostly Caucasian populations find that the prevalence (number of patients with the disease) of SLE is between 14.6 and 50.8 per 100,000, with an incidence rate per year (number of new cases annually) of 1.8 to 7.6 per 100,000. Nearly 60,000 individuals are paid members of the two largest lupus support organizations (the majority have lupus), which indicates a considerable amount of networking on the part of patients with the disease. A recent Lupus Foundation of America survey suggested that the prevalence of SLE may be as high as 2 million in the United States.

In Europe, some of the socialized medical systems compile diagnosis-based data banks. Among overwhelmingly Caucasian populations in Western Europe and Scandinavia, several surveys show a prevalence ranging from 12.5 to 39 per 100,000.

## WHY IT IS IMPORTANT TO KNOW THE PREVALENCE OF SLE

In order to find a cure for lupus, a pharmaceutical company must know its targeted population. Financial incentives to pour millions of dollars into research and development toward finding a new therapy are lacking if the numbers are too small. To deal with this problem of prevalence, the federal government created a category of ''orphan diseases,'' which are defined as disorders afflicting fewer than 200,000 people. If drug companies put research funds into these orphan diseases, they are given special financial incentives and a ''fast track'' for developing new remedies. As of this writing, SLE is considered an orphan disease.

## AGE OF ONSET

Lupus has been recorded in individuals at birth (neonatal lupus) and has been diagnosed in some people as old as 89. Nevertheless, 80 percent of those afflicted with SLE develop it between the ages of 15 and 45. Neonatal lupus is an acquired subset of the disease usually limited to children of mothers who carry a specific autoantibody (an antibody that reacts against the body's own tissues) called the anti-Ro (or SSA) antibody, which will be discussed in Chapter 30. This is one of the autoantibodies that crosses the placenta. For example, the skin rash of neonatal lupus is a self-limited process that disappears during the first year of life because the mother's antibody gets ''used up'' and the baby cannot make more of it. Children may develop SLE between the age of 3 and the onset of puberty. This form of lupus is usually a severe, organ-threatening disease but fortunately accounts for less than 5 percent of all lupus cases. The onset of lupus after age 45 or after menopause is uncommon, and a diagnosis of lupus past the age of 70 is extremely unusual. Late-onset lupus is generally mild and does not threaten organ systems, but it can be mistaken for rheumatoid arthritis, Sjögren's syndrome, or polymyalgia rheumatica (see Chapters 22 and 23 for a discussion of these conditions).

## SEX OF SLE PATIENTS

In children and in adults over the age of 50, the incidence of lupus demonstrates only a slight female predominance; however between the ages of 15 and 45, close to 90 percent of diagnosed patients are women. The reasons for this are discussed

**Table 3.** *Sex Ratios at Age of Onset or at First Diagnosis of SLE*

| Age | Female-to-Male Ratio |
|---|---|
| 0–4 | 1.4:1 |
| 5–9 | 2.3:1 |
| 10–14 | 5.8:1 |
| 15–19 | 5.4:1 |
| 20–29 | 7.5:1 |
| 30–39 | 8.1:1 |
| 40–49 | 5.2:1 |
| 50–59 | 3.9:1 |
| 60–69 | 2.2:1 |

in Chapter 17. Overall, 80 to 92 percent of all Americans with SLE are women. The percentages are less for discoid lupus, where 70 to 80 percent are women, and for drug-induced lupus, which occurs equally in males and females. In light of these statistics, lupus has been called a ''women's disease.'' To view the prevalence of lupus in men and women by ages, Table 3 summarizes some of the studies relating to sex and incidence.

## RACE AND GEOGRAPHY

The *incidence* of a disease is defined as the number of new cases per time period (e.g., year), whereas *prevalence* denotes the number of sufferers in the population. In the United States, African Americans, Latinos, and Asians have a greater incidence of SLE than Caucasians. The prevalence among African American women was estimated by Kaiser-Permanente to be 286 per 100,000 in San Francisco. A Hawaiian study showed that Asian women had three times the prevalence rate of SLE as compared with Caucasian women. American Indians seem to have the highest prevalence of lupus ever reported, but the numbers surveyed were too small to confirm this trend.

Within these broad groupings, geography and racial characteristics may influence the prevalence of lupus. For example, lupus is very rare on the African continent in comparison with the prevalence figures we see in the United States. It is much more common in the Philippines and in China than it is in Japan, and Sioux Indians have ten times the incidence of lupus as compared to other American Indian tribes. Asians more often tend to have severe organ-threatening disease compared with other demographic groupings, closely followed by African American males.

## WHY DO PEOPLE GET LUPUS?

Lupus results when a specific predisposing set of genes is exposed to the right combination of environmental elements, infectious agents, lupus-inducing drugs, excessive ultraviolet light, physical trauma, emotional stress, or other factors. The next few chapters detail the circumstances that make certain populations more susceptible to the disorder than others.

# Part II
# INFLAMMATION AND IMMUNITY

Part I defined and classified lupus, explored the historical context of this disease, and reviewed the populations lupus afflicts. The next two parts look at how it damages body tissue and why it occurs. Scientifically speaking, this is the most difficult part of the book, because we tackle complex immunologic concepts and discuss how inflammation takes place. Tables and summaries are provided throughout to assist the reader. Feel free to skip this section or skim it. First, we turn to the workings of the normal immune and inflammatory response so that the abnormal responses observed in lupus will be better understood.

# 5
# *The Body's Protection Plan*

Inflammatory and immune responses account for many of the symptoms observed in systemic lupus. This chapter reviews concepts of immunity and inflammation; the following chapters discuss how these concepts apply to rheumatic diseases.

## WHAT ARE THE COMPONENTS OF THE NORMAL INFLAMMATORY AND IMMUNE SYSTEM?

The body is always on the lookout for foreign substances that may pose a threat to its intricate workings. Its monitoring system consists of blood and tissue components, including certain proteins and blood cells that travel back and forth between blood and tissues.

### Blood Components

A 150-pound (70-kilogram) person has about 6 liters of blood, which contains several components. These include *red blood cells,* called *erythrocytes,* which are responsible for carrying and exchanging oxygen. If a person has a low count of red blood cells, she is suffering from anemia. *White blood cells,* call *leukocytes,* constitute the body's main defense system. Other blood components are *platelets,* which clot blood, and *plasma,* which includes serum. Plasma makes up most of our blood volume. It contains many proteins and other substances being carried to different parts of the body, including clotting factors that are not present in serum.

*White blood cells* play a central role in inflammation. Five types of white blood cells have heen identified by scientists; all are relevant to lupus. These include the following:

*Polymorphonuclear cells.* These cells are also called *neutrophils* or *granulocytes* and, like all other blood components, they are made in our bone marrow (blood-making parts of our bone in the pelvis and sternum). After being produced, they circulate in the blood for a few days and then pass into tissues. Some 50 to 70 percent of our circulating white cells are neutrophils.

*Eosinophils.* These white blood cells make up 0 to 5 percent of all our white blood cells. Their life cycle is similar to that of granulocytes. Eosinophils are involved in allergic responses.

*Basophils.* These cells do not have a clearly defined function and constitute less than 1 percent of our white blood cells. Tissue-based basophils are termed *mast cells.* These specialized cells combat parasitic or fungal invasion. They also play a role in allergy.

*Lymphocytes.* These make up 20 to 45 percent of our white blood cells and are the gatekeepers of our immune responses. Produced in the bone marrow, they migrate constantly between blood and tissue and can survive as long as 20 years. Lymphocytes can be T (thymus-derived) or B (derived from the mythical ''Bursa of Fabricius'') cells.

*Monocytes.* These cells represent about 5 percent of our circulating blood cells. They are the circulating blood component of what is called the ''*monocyte-macrophage*'' network because these cells are responsible for processing foreign materials (antigens) and the destroying cells and tissue debris that are by products of inflammation. In circulating blood, these cells are called monocytes; macrophages can also be present in blood, but they are mostly in tissues (see Table 4, Figures 1 and 3).

**Table 4.** *Circulating Components of Whole Blood Important to the Immune System*

- Red blood cells
- Platelets
- White blood cells (leukocytes)
  - Basophils (called mast cells in tissues)
  - Eosinophils
  - Polymorphonuclear cells (granulocytes, neutrophils)
  - Monocytes (called macrophages in tissues)
  - Lymphocytes
    - T cells
      - CD4 cells (helper)
      - CD8 cells (suppressor)
    - Natural killer cells
    - B cells
- Plasma (includes serum)
  - Albumin
  - Globulin
    - Alpha globulins
    - Beta globulins (includes complement)
    - Gamma globulins (includes immunoglobulins, listed below)
      - IgG
      - IgA
      - IgM
      - IgD
      - IgE
  - Cytokines

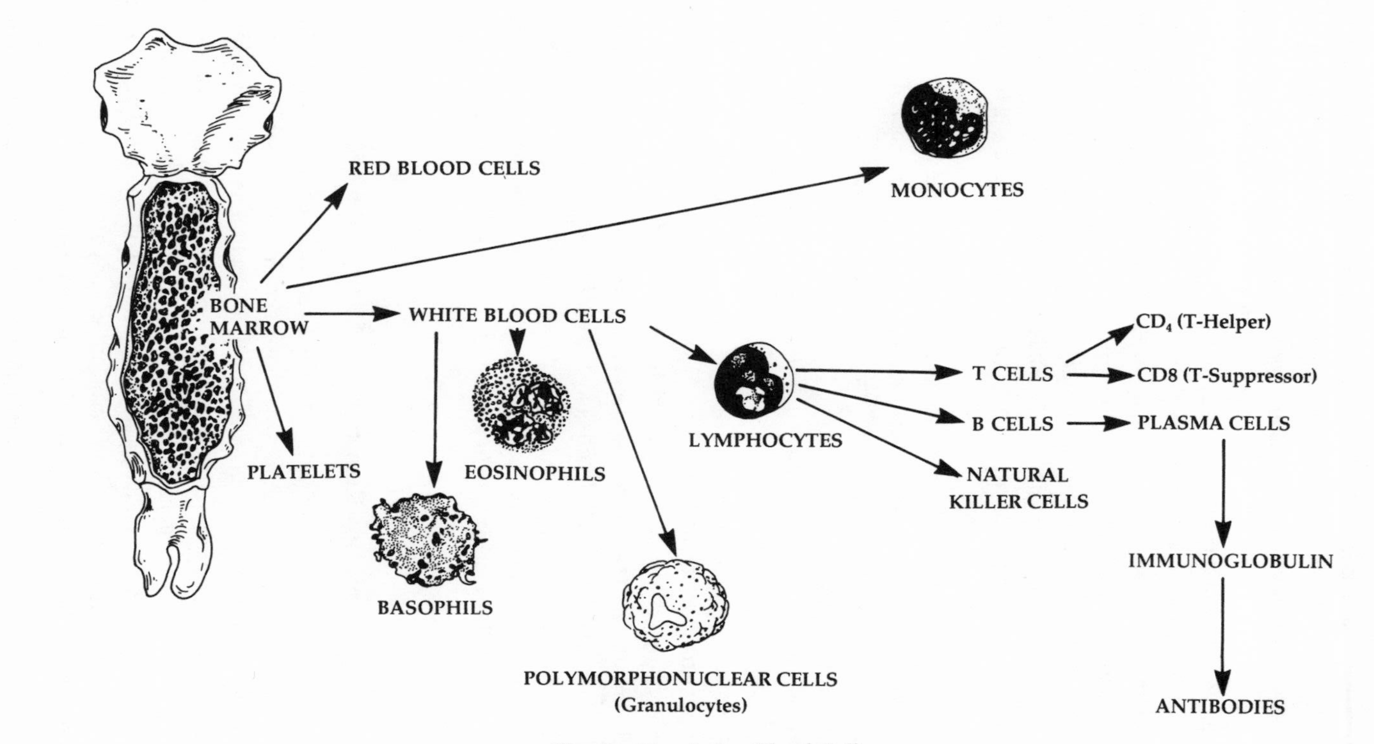


**Fig. 1.** *Circulating Blood Cells*

## Lymphoid Tissue and the Thymus

Lymphoid tissue is a key part of the immune system and represents up to 3 percent of a person's body weight. It includes our *lymph nodes* (or lymph glands), the *circulating lymphocytes,* and fixed *lymphoid tissue* (i.e., spleen). A 150-pound person has $10^{12}$, or 100 billion lymphocytes. They are widely distributed throughout the body and consist of long- and short-lived populations.

Bone marrow is the source of primitive ancestors of the *T* and *B lymphocytes.* These precursors migrate to the thymus, a gland just below the neck, which processes them into immunologically competent and knowledgeable *T cells.* These T cells provide cellular immunity and are the body's memory cells. About 70 percent of the lymphocytes are T cells. They remember what is foreign, go on to alert the body when a person reencounters a foreign substance, and formulate a response that protects the body.

Blood is carried to tissues by the arteries, and returns to the heart through the veins. Blood components, cellular waste and debris, and other materials can also return by another system—a chain of *lymph nodes* that starts in our toes and fingers and ends up in the chest area.

Lymphoid tissue contains T cells, *B cells,* and *natural killer cells.* B cells, which make up 10 to 15 percent of the lymphocytes, produce antibodies that eliminate what is foreign. Natural killer cells destroy targeted cells without having been sensitized to them in the past.

There are various types of T cells, which are identified by their surface markings and appearance. These types are labeled by the cumbersome term *cluster-determined,* or *CD.* Nearly all T cells have markers associated with *CD3. CD4* cells are those that "help" or promote immune responses, while *CD8* cells usually "suppress" or block the immune response. Approximately 50 percent of T cells have the CD4 marker and 20 percent the CD8 marker. Other markers are also present.

## Our Antibody Response: The Gamma Globulins

When you were growing up, there may have been an occasion when your pediatrician gave you gamma globulin shots to minimize certain infections that were going around. A type of gamma globulin, called immunoglobulin, is responsible for our antibody response. In response to an antigen, or foreign substance, our bodies produce antibodies. With appropriate signaling by T cells, B cells transform themselves into *plasma cells.* Plasma cells make immunoglobulins. These gamma globulins circulate in the plasma and protect the body from infection and other foreign material. There are five types of immunoglobulins:

*IgG* (*immunoglobulin G*) is the major antibody of plasma and the most important part of our antibody response. Most autoimmune diseases are characterized by IgG autoantibodies.

*IgM* is initially produced to fight antigens but soon decreases and allows IgG to take over. It plays an important but secondary role in autoimmunity.

*IgA* is the major antibody of external secretions (tears, gastrointestinal tract secretions, and respiratory tract secretions). It is important in Sjögren's syndrome (a combination of dry eyes, dry mouth, and arthritis seen in many lupus patients) and autoimmune diseases of the bowel (ulcerative colitis and Crohn's).

*IgD* is poorly understood but has a role in helping B cells recognize antigen.

*IgE* binds to mast cells and mediates allergic reactions.

This categorization is summarized in Table 4 and Figure 1.

### And Finally, Cytokines and Complement

*Cytokines* are hormonelike substances that promote various activities in the body, but in lupus, their functions are altered. Cytokines play a role in the growth and development of cells, and include various interleukins and interferons. For example, interleukin-1 has many actions. Secreted during the course of an immune response, it exerts effects by binding to receptors on the cell surface. Interleukin-1 can stimulate T cells to make interleukin-2, trigger the liver to make chemicals that perpetuate inflammation, allow certain cells to proliferate, and promote the production of growth factors which, in turn, make more white blood cells and other growth factors, thus amplifying or "gearing up" the immune system.

*Complement* refers to a group of 28 plasma proteins whose interactions clear away immune complexes (antigens mixed with antibodies) and kill bacteria. They are consumed (serum levels decrease) during inflammation, and low complement levels are an important indicator of lupus activity.

## THE INFLAMMATORY PROCESS

We have described the key fighters in the body's defensive army against immunologic and inflammatory attack. We can imagine them as a highly disciplined force, each member of which carries out a specialized task in the course of battling against foreign invaders. Neutrophils, lymphocytes, and macrophage-monocytes are all involved in the body's inflammatory and immune process in critical but distinct ways.

### Neutrophils and Inflammation

In healthy people, neutrophils have only one known function: they kill foreign invaders such as bacteria. If the level of neutrophils in the blood is low, we know this decreases our ability to fight infection and increases our risk of contracting it.

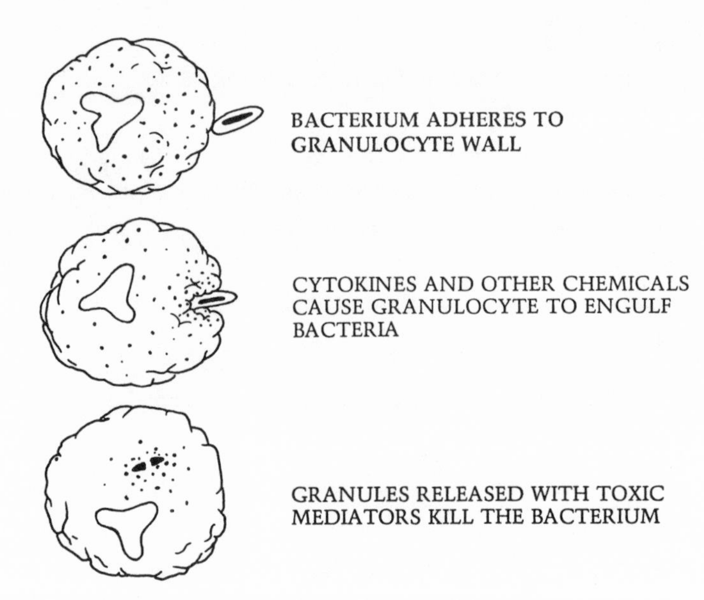


**Fig. 2.** *How Granulocytes Kill Bacteria*

In rheumatic diseases such as gout, neutrophils can ingest or swallow immune complexes and crystalline material. Neutrophils are part of the acute (early, initial) inflammatory process, whereas lymphocytes are part of a chronic (later, ongoing) inflammatory process.

The process by which neutrophils kill foreign material occurs in several stages, which can be visualized in Figure 2. First, try to imagine neutrophils or antibodies as guns and cytokines or complement as bullets, which cause tissue destruction. By adhering to the surface of veins and emigrating through them, neutrophils turn on a system of attractants that includes activated complement and cytokines as well as toxic mediators, which are important for cell destruction and inflammation. These mediators generate chemicals that can be suppressed by lupus medicines such as steroids and nonsteroidal anti-inflammatories (e.g., Advil, Naprosyn). The end result is the coating of bacteria with IgG and activated complement, which then adheres to the neutrophil. Finally, the neutrophil discharges its granules, thus completing the killing process.

## The Monocyte-Macrophage and Antigen-Presenting System

The monocyte-macrophage network is an important member of the immune surveillance force; it is central because regulation of this network is what goes awry in autoimmune disease. The monocyte-macrophage network has several functions: it destroys microorganisms and tissue debris that result from inflammation; it clears dead and dying red cells, denatured plasma proteins, and micro-

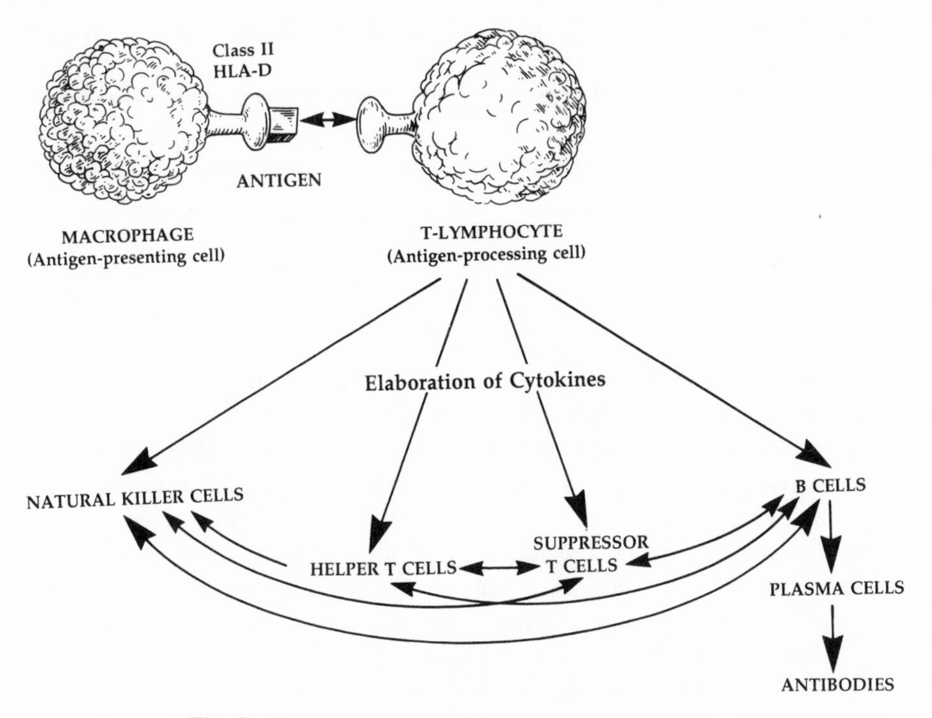


**Fig. 3.** *How Antigens Stimulate Antibody Production*

organisms from the blood; it plays a role in the recognition of foreign substances; and it also promotes the secretion of cytokines.

Antigens, or foreign materials, do not generally activate T cells directly. They are usually presented to the T cells by macrophages. Antigens are present on the macrophage surface, but in order to respond to their presence, the T cell must recognize a code on the surface of the macrophage, called the HLA class II (or D) determinant. The HLA (or human leukocyte antigen) system is responsible for recognizing antigen. See Chapter 7 for a review of this system. This system acts in combination with a T-cell surface marker, which can then activate T cells. Class II determinants recognize surface antigens for CD4, or helper cells; class I (HLA-A, B, or C) determinants recognize markers for CD8 or suppressor cells. Figure 3 illustrates these complex interactions.

## Activation of Lymphocytes

The T lymphocytes are made in the bone marrow and processed in the thymus. When they leave the thymus, they are able to respond selectively to environmental stimulation. Upon exposure to a foreign body or antigen and after a series of steps, T cells transform; they become larger in appearance and start to divide. This occurs as a result of the production of the cytokine interleukin-2 (also called

T-cell growth factor). T cells then differentiate into helper cells, suppressor cells, effector cells (which make cytokines), and cytotoxic or killer cells, and they promote the production of B cells. Some B cells become plasma cells and make immunoglobulin, or antibody. A small number of T cells live for many years and act as memory cells for the immune system. They are capable of initiating effective and rapid immunologic responses if the body is re-exposed to the antigen.

## Summing Up

Whole blood consists of red blood cells, white blood cells, platelets, and plasma. There are five types of white blood cells. These include the neutrophils, which are important in acute inflammation; lymphocytes, which help regulate chronic inflammatory processes; and monocytes, which are responsible for helping the body recognize foreign material. All these cells are derived from the bone marrow. In a normal immune response, these elements all work together. Lymphocytes migrate to the thymus, where they are ultimately recognized as T cells or B cells. The T cells read antigenic signals present in monocytes or macrophages and are thus able to promote or turn off inflammation and the killing of specific cells. Some B cells transform themselves into plasma cells that make immunoglobulin, which then circulates in the plasma. Immunoglobulins G, A, M, D, and E also help to destroy foreign materials. All these processes are promoted and amplified by cytokines, complement, and other mediators that constitute our normal immune surveillance network.

# 6

# *The Enemy Is Our Cells*

Chapter 5 reviewed the normal inflammatory and immune response network. Several features unique to lupus and other autoimmune processes alter this system to produce tissue injury that is not observed by the body's normal immunological surveillance system. These features will be summarized briefly here in order to help us understand the many antibodies that play an important role in lupus.

Lupus results when genetically susceptible individuals are exposed to certain environmental factors, and in my opinion, only 10 percent of those who carry lupus genes will ever develop the disease. These environmental factors create a setting where things happen that normally shouldn't. *Neutrophils* (the white blood cells responsible for mediating acute inflammation) can increase inflammation in the body of lupus sufferers because of the way their blood plasma interacts with cytokines, complement, and adhesion molecules (the chemicals that draw cells closer to the site of inflammation). *Lymphocytes,* the white blood cells responsible for chronic inflammation, also have their function altered in lupus. The T-helper cells become more active, and the body becomes less responsive to T-suppressor cells. Natural killer lymphocytes promote inflammation and are not able to suppress or contain it. As a result, the body's system of tolerance is disrupted so much that B cells are signaled to make antibodies to the patient's own tissues, which are called *autoantibodies.* In other words, the normal immune surveillance system is altered in lupus, resulting in accelerated inflammatory responses and autoantibody formation; the autoantibodies, in turn, attack the body's own cells and tissues. It is as if our body's police force found itself unable to tolerate healthy, law-abiding cells and schemed to undermine them.

## AUTOANTIBODIES GALORE!

Autoantibodies are the hallmark of lupus. They represent antibodies to the body's own tissue—to parts of the cells or the cells themselves. In addition to having a positive result on an antinuclear antibody (ANA) test, the typical SLE patient has at least one or two other autoantibodies. These antibodies distinguish lupus patients from others without the disease, since few healthy people have significant levels of them.

Sixteen important autoantibodies in lupus are described and defined in this chapter. In Chapter 11, the clinical importance of some of these autoantibodies is discussed in more detail; there, we will look at the blood tests your doctor will order so that your disease can be diagnosed and treated. In other words, this section defines and categorizes autoantibodies.

## WHERE DO AUTOANTIBODIES COME FROM?

Autoantibodies are triggered by antigens in the environment (e.g., foods, dyes, silicone in breast implants), autoantigens (self-driven), and *idiotypes.* Idiotypes are regions of immunoglobulin that are recognized as being foreign, or antigenic. They promote the formation of *antiidiotypic antibodies.*

Even though immunologists know that B cells are critical in the formation of autoantibodies, several important issues regarding autoantibodies remain unresolved. Namely, which B cells can make autoantibodies? And what is the defect in immune response that permits the expression of pathologic or injurious autoantibodies?

In nature, autoantibodies are often produced that do not have any clinical or pathologic significance (mostly IgM antibodies). In lupus, IgG autoantibodies often correlate with disease activity. Autoantibody responses can be general or on occasion quite specific. Both phenomena are observed in lupus.

## WHAT KINDS OF AUTOANTIBODIES ARE THERE?

The autoantibodies important in lupus can be broken down into four categories. These include *antibodies that form against materials in the nucleus or center of the cell,* such as antinuclear antibody, anti–DNA, anti-Sm, anti-RNP, and antihistone antibody; *antibodies that form against cytoplasm or cell surface components,* such as anti-Ro (SSA), anti-La (SSB), antiphospholipid, and antiribosomal P antibodies; *antibodies to different types of cells,* such as red blood cells, white blood cells, platelets, or nerve cells; and *antibodies that form against circulating antigens,* such as rheumatoid factors and circulating immune complexes.

If you are a lupus patient, your doctor is likely at some time to test your blood for many if not most of these antibodies. Therefore the following paragraphs endeavor to explain what they are. Doctors have been accused of creating their own vocabulary in order to make themselves indispensable. Consider this chapter a foreign language class. The next time Dr. Jones excitedly tells you that your anti-Ro (SSA) antibody is highly positive, ask him or her if it was measured using immunodiffusion or by ELISA!

## ANTIBODIES TO NUCLEAR CELLULAR COMPONENTS

Since its description in 1957, the *antinuclear antibody,* or ANA (antibody to the cell nucleus) has become the most widely known autoantibody in SLE. It is hard to imagine having lupus without it, although a person can have a positive ANA test without having systemic lupus. For many years, the test for ANA was conducted using animal cells. Human serum was placed over kidney or liver cells of the mouse, rat, or hamster, and if ANA was present, human antibody would attach to the animal cell's nucleus. Fluorescent staining of the antibody bound to animal cell nuclei was used to document these findings. Now, however, testing involves human cells in place of animal cells; as a result, the test more accurately predicts the presence of lupus. As recently as the late 1980s, 10 percent of lupus patients had false-negative results (in other words, people with lupus had a negative ANA test) and might have been told they did not have the disease. The rate of negatives at present is as low as 2 percent; however, more people now have positive ANAs without evidence for lupus.

ANA tests are analyzed according to the amount of antibody present and the pattern seen in cells recognized by antibodies in the sample. Although this is a crude measure, the amount of antibody can suggest the degree of seriousness of the disease. The patterns of antibody are classified as follows:

1. The *homogeneous* pattern is seen primarily in SLE but also in other illnesses as well as in older adults. This pattern indicates the presence of antibodies to chromatin (part of the chromosomes), histone, and deoxynucleoprotein.
2. The *peripheral* or *rimed* patterns are seen primarily in SLE and represent antibodies to DNA (see below). Often, the concurrence of a homogenous pattern makes the peripheral pattern undetectable, since a stain that covers all the nucleus may include its rim or margin.
3. The *speckled* pattern is seen in SLE, numerous other autoimmune diseases, and in some healthy individuals who show low amounts of antibody. "Speckled" suggests a spotty uptake of the fluorescent stain.
4. *Nucleolar* (a part of the nucleus) patterns are not often seen in SLE and suggest scleroderma.
5. *Centromere* patterns detect the central part of the chromosome. They are rarely seen in lupus and their presence suggests a form of scleroderma called the CREST syndrome.

As many as 10 million Americans may receive a positive ANA test result, but fewer than a million probably have SLE. Positive ANAs (even in large amounts) can be found in healthy relatives of SLE patients, patients with other autoimmune diseases, and in normal people, especially those over the age of 60. In addition, infections with certain viruses and bacteria can stimulate the production of ANA.

*Deoxyribonucleic acid,* or *DNA,* is a molecule located in the control center of each of our cells and is responsible for the production of all the body's proteins. Antibodies to single-stranded DNA are found in many normal individuals. However, the presence of *antibodies to double-stranded DNA* may suggest a serious form of SLE; these antibodies, if positively charged, may damage tissue directly. Throughout, I will refer to these antibodies as *anti-DNA*. Approximately half of SLE patients possess anti–DNA. By tracking their specific levels, I can assess my patient's response to therapy. More than 90 percent of patients with anti–DNA have SLE, especially those with serious organ-threatening (e.g., kidney) disease.

*Anti-Sm* stands for a Mrs. Smith, in whom it was first described. Antibodies to the Sm antigen are very specific for lupus (present in 20 to 30 percent of those with SLE) and are rarely observed in any other disease. These antibodies interfere with the ability to transcribe RNA (ribonucleic acid) from DNA.

The antibody directed against ribonuclear protein, *anti-RNP,* is essential for the diagnosis of mixed connective tissue disease, a lupus look-alike. However, it is not specific to this disease, since 20 to 30 percent of lupus patients and a small number with scleroderma or rheumatoid arthritis also have the antibody. Anti-RNP interferes with the ability of RNA to bind in the cytoplasm of cells.

Histones are structural proteins in the cell nucleus. They can be autoantigens and are observed in SLE, rheumatoid arthritis, certain cancers, and liver diseases. Histones are thought to be responsible for the LE cell phenomenon, the first immune abnormality reported in lupus. *Antihistone antibodies* are of particular interest because they are present in 95 percent of patients with drug-induced lupus.

## ANTIBODIES TO CYTOPLASMIC COMPONENTS

Four autoantibodies to cytoplasm (if a cell looks like an egg, it is the egg white) are important in SLE and are discussed here. The true prevalence of many of these antibodies, however, is not known, since the levels vary according to the methods of detection employed. Although the technology is not relevant to this discussion, it is important to know that older, less sensitive methods such as immunodiffusion detect fewer positive patients than the newer ELISA and immunoblotting tests. But whereas the older tests rarely if ever gave false-positive readings, newer methods of evaluation occasionally produce misleading results. These methods nevertheless detect 30 to 50 percent more patients with the autoantibody than traditional testing. The percentages listed below for anti-Sm, anti-RNP, anti-Ro, and anti-La were derived from older methods.

*Antiphospholipid antibodies* react against phospholipids, which are components of the cell membrane. There are several antiphospholipid antibodies, the most important of which is anticardiolipin. These antibodies are frequently seen with the lupus anticoagulant and are discussed in detail in Chapter 21. Antiphospholipid antibodies are present in about one-third of patients with SLE and are less frequently found in other autoimmune diseases.

*Anti-Ro* (*SSA*) is present in most patients with Sjögren's syndrome (dry eyes, dry mouth, and arthritis) and 20 to 30 percent of those with SLE. As one of the few autoantibodies that crosses the placenta, it may induce neonatal lupus and congenital heart block. Anti-Ro can impart increased sun sensitivity to its carriers and is seen in nearly all patients with a skin disorder called subacute cutaneous lupus erythematosus. It is associated immunogenetically with HLA-DR2 and HLA-DR3. Anti-Ro may interfere with the cell's ability to process RNA.

*Anti-La* (*SSB*) is almost always seen with anti-Ro. Rarely present by itself, it coexists with anti-Ro 40 percent of the time and may make anti-Ro less pathogenic, or dangerous. This statement especially applies to kidney disease. La may function as a waystation on the road to where RNA transcripts are carried from the nucleus to the cytoplasm.

*Antiribosomal P* antibodies are observed in 20 percent of known patients with SLE. Found only in the cytoplasm, antiribosomal P may correlate with psychotic behavior or depression in lupus and its levels may decrease with response to therapy.

## ANTIBODIES TO CELLS

Lupus patients can have antibodies to red blood cells, white blood cells, platelets, and nerve cells.

*Antierythrocyte antibodies* are directed against the surface of red blood cells. One lupus patient in 10 develops autoimmune hemolytic anemia, or the destruction of red blood cells due to antierythrocyte antibodies. The true prevalence of these antibodies is not known, but there are probably two to three lupus patients with this antibody for every one who becomes seriously anemic.

*Antineutrophil antibodies* are occasionally observed. Antibodies that act against the cytoplasmic component of neutrophils, called *ANCA* (*antineutrophil cytoplasmic antibodies*), indicate the presence of one of two types of non-lupus vasculitis (inflammation of the blood vessels) called Wegener's granulomatosis (which is characterized by the c-ANCA subset) and polyarteritis nodosa. The p-ANCA subset, which characterizes polyarteritis, is also positive in 20 percent of those with SLE.

Another white blood cell antibody called *antilymphocyte antibody* is present in most lupus patients. It is a natural autoantibody; large amounts indicate greater severity of the disease, and it is responsible for the low white blood cell counts seen in many lupus patients. Most antilymphocyte antibodies are IgM antibodies that coat the surface of lymphocytes and result in the depletion of T cells.

*Antiplatelet antibodies* are present in approximately 15 percent of lupus patients and account for most thrombocytopenia (a condition in which platelet counts drop below 100,000 per cubic millimeter) seen in lupus patients. Although there are many other reasons for thrombocytopenia (they are discussed in detail in Chapter 20), the majority of patients have platelets coated with IgG, which assures their premature destruction. Unfortunately, the methods for detecting antiplatelet antibodies are technically difficult and not always reproducible. Antiplatelet antibodies are often seen in association with antiphospholipid antibodies.

*Antineuronal antibodies* react to nerve cell membranes, or coverings. Nerve cells, or neurons, look like a series of electrical wires and are responsible for storing and transmitting nerve responses throughout the body. Autoantibody reactivity to nerve cell components is not observed in healthy people, but if tested correctly can be detected in 5 percent of patients with rheumatoid arthritis, 5 to 20 percent of all lupus patients, and in up to 90 percent of patients with active central nervous system inflammation from lupus or autoimmune nervous disorders. Although blood levels may indicate the presence of central nervous system lupus, spinal fluid is a more reliable source than serum for measuring neuronal antibodies.

## ANTIBODIES THAT FORM AGAINST CIRCULATING ANTIGENS

Remember that antigens are foreign materials to which cells react. Sometimes the body makes antibodies to antigens expressed by its own cells.

In rheumatoid arthritis, it is probable that antigenic material to the synovium (the lining of the joint) induces an antibody response. The body then makes antibody to this antigen-antibody complex which is called *rheumatoid factor*. Rheumatoid factor, in turn, may release a chain of biochemical events that contributes to joint and cartilage destruction. As many as one-third of lupus patients may have rheumatoid factor. Rheumatoid factor in lupus can exacerbate the inflammation of joints but may be protective of other tissues such as the kidneys.

*Circulating immune complexes* are the combination of antibody and antigen circulating in the bloodstream. These complexes contain everything from rheumatoid factor to complement mixed with antigen and immunoglobulins. The true incidence of circulating immune complexes in lupus is unknown, since the variety of methods capable of assessing them detect only certain types of complexes. Circulating immune complexes can activate complement which, in turn, can promote inflammation. In lupus, some complexes cannot be cleared from the body by the monocyte-macrophage system, and when they settle in tissues either directly or indirectly, they induce inflammation.

Table 5 summarizes these antibodies and compares their presence in normal individuals and lupus patients.

**Table 5.** *Important Autoantibodies and Antibodies in Lupus*

| Autoantibody | Antibody to | Percent in Lupus | Percent in Normals | Lupus Specificity |
|---|---|---|---|---|
| Antinuclear | Nucleus | 98 | 5–10 | Fair |
| Anti–DNA | Nucleus | 50 | <1 | Excellent |
| Antihistone | Nucleus | 50 | 1–3 | Fair |
| Anti-Sm | Nucleus | 25 | <1 | Excellent |
| Anti-RNP | Nucleus | 25 | <1 | Fair |
| Antiphospholipid | Membrane | 33 | 5 | Fair |
| Anti–Ro/SSA | Cytoplasm | 30 | <1 | Fair |
| Anti–La/SSB | Cytoplasm | 15 | <1 | Fair |
| Antiribosomal P | Cytoplasm | 20 | <1 | Good |
| Antierythrocyte | Red cells | 15–30 | <1 | Fair |
| ANCA | White cells | 20 | <1 | Poor |
| Antilymphocyte | White cells | Most | 20 | Poor |
| Antiplatelet | Platelets | 15–30 | <5 | Poor |
| Antineuronal | Nerve cells | 20 | <1 | Good |
| Rheumatoid factor | Ag-Ab* | 30 | 5–10 | Poor |
| Immune complexes | Ag-Ab* | Most | Varies | Poor |

* Ag-Ab is an abbreviation for antigen-antibody complexes. These immune complexes are elevated with many common bacterial and viral infections, not just with lupus.

## LESSONS LEARNED FROM ANIMALS

What is fascinating about animal research is that we have found similarities between the immune systems of animals and those of humans. Much of the information presented here was derived from animal research. In addition to affliciting nearly a million Americans, lupus is also found in many animals. It has been reported to occur spontaneously in dogs (including a presidential one—George Bush's Millie), cats, rabbits, rats, mice, guinea pigs, pigs, monkeys, goats, hamsters, and Aleutian minks. One of the best described is canine lupus, which is very similar to human lupus in its presentation and management. In the 1950s and 1960s, occasional research studies utilized guinea pigs and rabbits, but these approaches have been abandoned. One laboratory has extensive experience inducing lupus in cats with an antithyroid preparation, but this has not been adopted as a research tool by other investigators. More than 95 percent of animal lupus research studies involve mice with lupus.

## WHY SHOULD WE STUDY ANIMAL MODELS OF LUPUS?

Antivivisectionists loudly proclaim that animal research is inhumane and unnecessary. They believe that computer modeling and tissue-culture work rule out the need for animal studies. However, many of the advances in lupus over the last 30 years would not have been possible without animal studies, and thousands of human lives have been saved as a result of this work. The immune system of a mouse is remarkably similar to that of a human. As hard as we try, no satisfactory computer simulation of the mouse's or human's immune system exists, in large part because there's a lot we don't know about it.

The breakthroughs resulting from mouse work in SLE include proof that genetic factors are important determinants of autoimmune disease and have led to the identification of genes important in lupus. Animal research has also proved that lupus can be influenced by environmental and hormonal factors. Therapy has pushed forward because trials of multiple therapeutic interventions that never would have worked in humans have saved years of research and lots of misery for patients. Furthermore, many of the drugs we use to treat lupus (e.g., cyclophosphamide) were first shown to be effective in mice. There have even been strong suggestions from mouse work that "gene therapies" may become useful in the treatment of lupus.

As long as investigators stick to well-established guidelines that mandate humane and ethical environments for animal research, these efforts can save billions of dollars and a lot of unnecessary trial and error in humans while accelerating the pace for establishing the efficacy of new treatments. For example, since mice with lupus live 1 to 2 years and humans with lupus can live up to 100 years, the influence of different therapeutic or environmental interventions can be seen more easily in animals with lupus.

# Part III
# WHAT CAUSES LUPUS?

Now that we have seen how our complex immune system works, we can return to the question of what causes lupus. Researchers now believe that SLE probably results from multiple factors. It begins when certain genes predisposing an individual to lupus interact with environmental stimuli. These interactions result in immunologic responses that make autoantibodies (antibodies to one's self) and form immune complexes (antigens combined with antibodies). Certain autoantibodies and immune complexes are capable of causing the tissue damage typically seen in lupus.

This part briefly outlines the causes of lupus and the immunologic responses typically seen in lupus patients; subsequent chapters review these concepts in more detail.

## THE GENETICS OF LUPUS

Several different single genes increase the relative risk for lupus by increasing the body's ability to promote certain autoantibodies. These are called HLA (human leukocyte antigen) class II genes (there are class I, II, and III genes), and they are present on the surface of all cells that present foreign material, called antigens, to white blood cells, which are central in the body's immune system. A defect in HLA class III genes results in low complement levels (an important protein that plays a role in inflammation), which are commonly observed in SLE. Outside the HLA system, genes that help program the structure of immunoglobulin or the receptors on the surface of T cells are important as well. And as one might deduce from the much greater number of female SLE patients, sex hormones also play a role in the unique immunologic response of lupus patients. In fact, female sex hormones are more compatible with lupus activity than are male hormones.

## ENVIRONMENTAL FACTORS

Environmental factors such as ultraviolet light, certain prescription drugs, and some chemicals can promote lupus. They either act like antigens that react against the body or introduce new antigens to the immune system. Viruses and other microbes can also alter cellular DNA or RNA (the essential structural material of chromosomes) and make them respond as if they were antigens. For still unknown reasons, non-Caucasians are more susceptible to these events.

## ABNORMALITIES OF B LYMPHOCYTES

One-third of the human white blood cells are called lymphocytes. They are divided into B and T cells. B cells are the body's "humoral" response; their job is to produce immunoglobulins and ultimately antibodies. The B lymphocytes in lupus patients are overactive; that is, they produce abnormally large amounts of immunoglobulin and autoantibodies.

## ABNORMALITIES OF T LYMPHOCYTES

The T cells are our body's "memory" lymphocytes. They remember what is foreign (or antigenic) and signal us to respond to this stimulus. Different types of T cells have various functions: they suppress immune response (suppressor cells), promote the immune response (helper cells), destroy cells (natural killer cells), or promote chemicals (e.g., cytokines) that modulate or signal other immune cells to do certain things.

In SLE, the abnormalities observed include increased helper function, decreased suppressor function, an alteration of the lymphocytes that promote autoantibody formation, and increased B-cell responses.

## ABNORMALITIES OF IMMUNOGLOBULIN RESPONSE

B cells make immunoglobulins, which destroy foreign material and protect the body. In lupus, the regulation of this process goes awry when autoantibodies form or when immune complexes are deposited in tissue and produce inflammation. Autoantibodies that target different parts of cells, such as anti-DNA, anticardiolipin, or anti-Ro (SSA), are capable of damaging tissues directly.

## ABNORMALITIES IN DECREASING IMMUNE RESPONSES

In a healthy body, antigenic or foreign material combines with antibodies to form immune complexes that vary in size, shape, charge, and binding properties.

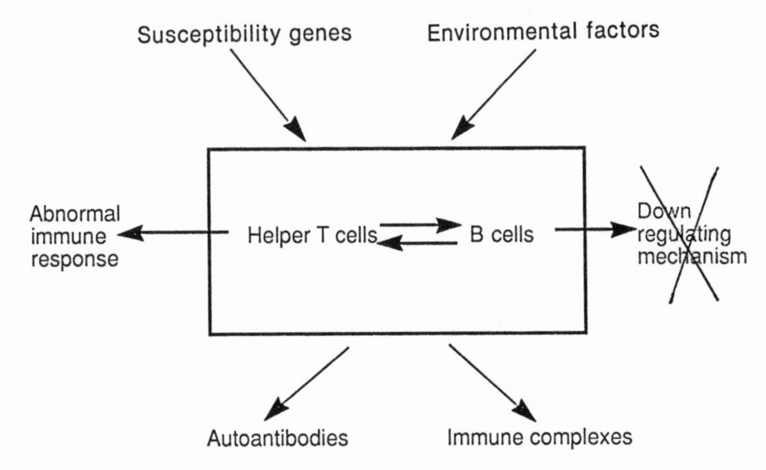


**Fig. 4.** *Factors Promoting the Development of SLE*

Circulating immune complexes are usually cleared and dispersed through a filtering system in which the spleen plays a prominent role. In SLE, this clearance is defective because certain important receptors have lost their binding ability and these complexes are unusually large or small in size or too plentiful.

Morecver, in the lupus patient, the body's ability to control inflammation is hampered because T-suppressor cells develop helper functions and natural killer cells promote B cells instead of killing certain invaders. Indeed, the body's system of "tolerance" (its ability to distinguish what is self from what is foreign) is altered.

## Summing Up

Genetic, environmental, T-cell, B-cell, and antibody factors combine to produce what we recognize as lupus. We still don't know which factors are the cart and which the horse. And we don't yet know why event A leads to event B. We do know that lupus results in alterations of immune regulation that cause the body to become sensitive to its own tissues. Figure 4 summarizes some of these concepts. In the next few chapters, we explore some of the factors that induce lupus.

# 7
# *The Genetic Connection*

Is lupus a genetic disorder? Does it run in families? How can it be passed on or be inherited? These guestions are commonly asked. The answer is not simple, but researchers now believe that various genes that predispose people to lupus are inherited, among them the *major histocompatibility complex,* or *MHC,* which includes the *human leukocyte antigen* (*HLA*) region, a specific area of the genes. Along with HLA, we also inherit T-cell receptor genes and thus T-cell receptors and other genes relevant to SLE, such as immunoglobulin genes. Each of us inherits a unique chemical signature, just as we inherit our blood type.

All this probably seems a little like alphabet soup, but in the next few pages we take a closer look at how the principles of genetics apply to our understanding of lupus.

## THE MAJOR HISTOCOMPATIBILITY COMPLEX: A GENE SYSTEM

Every human cell that contains a nucleus (or center) also contains twenty-three pairs of *chromosomes.* We inherit one member of every pair from each of our parents. These chromosomes store the genetic material responsible for determining whether you are a male or female, have red hair, are color blind, might develop cystic fibrosis, among other characteristics. In mapping human chromosomes, geneticists have referred to the "short" and "long" arms of the chromosomes, which have been numbered for convenience. On the short arm of the sixth chromosome lie a series of specific sites, called genetic markers, that determine what an individual's HLA system will look like. First described in the early 1970s, the HLA region contains genes that may predispose one to a remarkable number of diseases, especially rheumatic disorders.

HLA testing is very simple. Your physician needs only a few tubes of blood drawn from the arm. Its applications for lupus are not reliable yet, but once scientists find out a bit more about which types of disease are associated with which genes, HLA blood testing will come down to an issue of cost and become more widespread.

The HLA site consists of three well-defined and functionally distinct regions known as classes I, II, and III. Class I is expressed on all cells with a nucleus and is divided into A, B, and C subtypes. Class II is present on cells that are capable of presenting antigens (foreign material) to white blood cells and includes the D subtype. The D regions are further broken down into DP, DQ, and DR sub-regions, among others. Class III provides for the structural genes that produce a variety of substances important in lupus blood and tissues such as complement, tumor necrosis factor, and heat shock protein.

An HLA "marker" is given to patients based on the subtype they possess, A, B, C, or D. Numerous *alleles* ("designations" or " arrangements") can be found at the same marker or site; there are more than a hundred possible arrangements that are further subdivided. If this seems complicated, don't worry. This expanding area of knowledge often confuses the best immunologists and rheumatologists.

A combination of alleles at two HLA loci are name tags, or *haplotypes*. They can differ widely among various racial and ethnic groups. For example, HLA-B27 (the marker associated with a spinal disease known as ankylosing spondylitis) is found in 8 percent of Americans of European Caucasian ancestry but is very rare among African Americans.

The statistical chance that two particular haplotypes will occur together is about 2 percent (for example, A6 with B5). However, in certain rheumatic diseases, the chance that two alleles or arrangements will occur together may greatly exceed this.

Table 6 illustrates the classification of the HLA system. The labels themselves are not important for our purposes; what is important is that specific genes are inherited and these may predispose a person to lupus.

**Table 6.** *The HLA Region of the Major Histocompatibility Complex*

| |
|---|
| Class I (on all nucleated cells) |
| A markers |
| B markers |
| C markers |
| Class II (on all antigen-presenting cells) |
| DP markers |
| DQ markers |
| DR markers |
| Class III contains structural genes that produce chemicals important in lupus, such as |
| Complement |
| Hormones |
| Cytokines |

## WHY IS HLA IMPORTANT IN LUPUS PATIENTS?

What does all this mean for lupus? First of all, certain *subsets* of the disease are associated (in other words, they are often but not always found) with very *specific* HLA markers. For instance, neonatal lupus (lupus afflicting children at birth) is most often present in children who possess the A1, B8, DR3, and DQw52 haplotypes. Patients with discoid lupus tend to possess DR4 markers, and DR3 is present in those with a specific skin problem known as subacute cutaneous lupus. Sjögren's syndrome (dry eyes, dry mouth, and arthritis, which is seen in many patients with SLE) is associated with B8, DR3, and DRw52. DR2 and DR3 are more commonly observed in Caucasians of Western European descent than any other DR types. The presence of DQw1 correlates with certain autoantibodies such as anti-DNA, anti-Ro, and anti-La. "Null" or absent alleles can account for some of the deficiencies in blood complement levels that are frequently seen in SLE.

Even though we have yet to isolate a lupus "gene," certain genetic markers and other non-HLA genes correlate with specific lupus subsets and autoantibodies. Different sets or combinations of genes may be associated with as much as a twentyfold risk for developing SLE. Investigations into these areas are still in their infancy, and perhaps some as yet undescribed systems might turn out to be more important than those we have already identified as putting an individual at risk for developing lupus.

## WHAT IS THE RISK THAT A MEMBER OF A LUPUS PATIENT'S FAMILY WILL DEVELOP LUPUS?

If you have lupus, members of your immediate family, or first-degree relatives (brothers, sisters, parents, and children), are at a slightly increased risk for developing it too. Several surveys have estimated this risk at 10 percent for your daughter and 2 percent for your son. If you have lupus and have an identical twin, the chance that this sibling is similarly afflicted ranges from 26 to 70 percent. If your twin is fraternal, however, this figure is only 5 to 10 percent. Interestingly, the prevalence of SLE among all family members of lupus patients is 10 to 15 percent, while the chance of any of this group having autoimmune diseases (including lupus) is 20 to 30 percent. The most common other autoimmune disorders include autoimmune thyroiditis (also known as Graves' disease or Hashimoto's thyroiditis), rheumatoid arthritis, and scleroderma. The body may produce elevated levels of autoantibodies even though no specific immune disorder is present. For example, nearly half the first-degree relatives of my lupus patients may have a positive ANA blood test. However, ANA is only one of the four criteria that must be present for SLE to be diagnosed. Whether or not a positive ANA increases the chance of developing lupus isn't known; most ANA-positive family members of lupus patients feel well and have no symptoms.

Should children of lupus patients or family members be ''typed'' or screened for SLE? I don't recommend testing unless symptoms or signs point to some existing clinical problem. At this time, there is nothing we can offer those who carry a lupus ''gene'' or autoantibody and have no symptoms. In other words, by testing them, we would only make them anxious or worried. Only a small percentage of these individuals will ever develop the disease.

## THE FUTURE

It is very possible—indeed probable—that in the next 30 years we will be able to identify patients at risk for developing SLE by using HLA testing or other methods and that we'll then vaccinate them to prevent lupus. By that time, all lupus-causing genes will have been isolated and identified. Potentially this would allow us to manipulate these genes in patients with active SLE to turn off the disease process.

# 8
# *Environmental Villains*

Many of my newly diagnosed lupus patients examine everything they have done or experienced regarding travel, prescription medications, occupational activities, infections, and other factors in an effort to find a reason for their disease. In my experience, some individuals are convinced that they did something "wrong" and therefore became ill. When this occurs, soul searching represents a natural process that ultimately results in coming to terms with the diagnosis. Although the precise cause of lupus is not known in each case, there are indeed certain environmental factors that may occasionally play a role in initiating the disease or making it worse. How does this happen?

A few of these mechanisms are linked to environmental factors and may produce effects in a variety of ways. These include a virus, food, or a chemical acting as an antigen to which an antibody response is generated. Some of these agents in patients predisposed to lupus mimic antigens to which the body is sensitized, and the antibody response is wrongly directed against the environmental factor. Alternatively, an antigen or inciting factor such as ultraviolet sunlight can damage DNA and promote the production of anti–DNA as an immune response to the altered DNA.

Various medications may also play a role in inducing lupus. Drug-induced lupus is covered in the next chapter.

This chapter concerns itself with four types of potentially inciting agents: (1) chemical factors, such as chemical agents, metals, and toxins; (2) dietary factors, such as amino acids, fats, and caloric intake; (3) ultraviolet radiation; and (4) infectious agents, such as viruses and bacteria as well as their by-products.

## WHAT CHEMICAL FACTORS CAUSE LUPUS?

### Aromatic Amines

*Aromatic amines* are chemical agents that may induce or aggravate rheumatic disease. This class includes *hair-coloring solutions, hydrazines* (e.g., tobacco smoke), and *tartrazines* (e.g., food colorings or medication preservatives). Aromatic amines are broken down in the body by a process known as *acetylation.* An increased incidence of drug-induced lupus has been observed after exposure to aromatic amines in patients who are *slow acetylators,* or those who metabolize aromatic amines slowly. About half of all Americans are slow acetylators. The mechanism by which aromatic amines may induce an immunologic reaction is poorly understood, and only a small percentage of people exposed to these chemicals ever develop clinical immune disease.

*Hair-coloring solutions* containing aromatic amines, specifically paraphenylenediamine, can reproduce features of autoimmune disease in experimental animals. Several large-scale epidemiologic surveys that that have tried to find out whether aromatic amines induce lupus or cancer have yielded conflicting results. Do I advise my lupus patients to avoid hair dyes? No, since I have not seen any patients who reported a flareup because they used a hair-care product; also, they already have the disease when they visit me.

*Hydrazines* are present in hydralazine, a blood pressure medication known to induce lupus. In addition, these substances are found in a variety of compounds used in agriculture and industry and occur naturally in tobacco smoke and mushrooms. A single published report tells of a pharmaceutical worker who was occupationally exposed to hydrazines, developed lupus, and had reproducible symptoms and signs upon repeated exposure.

*Tartrazines* are preservatives found in certain food dyes (such as FD&C yellow No. 5) and in some medicine tablets. Occasional well-documented reports of tartrazine-induced lupus have appeared.

## Silica and Silicone

One of the most ubiquitous elements in nature, silicon, has been the focus of numerous studies. Nearly 50 years have elapsed since the initial observations that sandblasters exposed to *silica dust* may develop an autoimmune type of reaction characterized by lung nodules and scarring as well as autoimmune-mediated lesions in the kidney.

The injection of *silicone,* a synthetic liquid form of silicon, under the skin has been similarly associated with autoimmune reactions. Silicone can be broken down into silica in the body, and a few women who have undergone breast augmentation with encapsulated silicone gel implants have developed a rheumatic disease that looks clinically like scleroderma and to a lesser extent like lupus, although the risk is probably only one in several thousand.

An allergic type of reaction to silicone called *silicone adjuvant arthropathy* has also been seen. Patients experience symptoms of arthritis and feel muscle aches, but this disappears when the silicone implants are removed. This reaction tends to occur more often in women who have had significant leakage or rupture of the implant sacs. Nearly a hundred manufacturers made breast implants between 1960 and 1990 using a variety of silicone compounds that had widely divergent features and complications. Although the potential risks of different implants are still being studied, autoimmune or allergic reactions are uncommon and are found in only a small percentage of implant patients. On the other hand, hard silicone has been used for medical purposes for over 100 years (e.g., artificial joints, penile prostheses) with few problems, except for rare localized allergic reactions.

### Other Chemicals

Some chemicals can produce lupus-like symptoms as part of other diseases. *Scleroderma* is a first cousin of lupus, and many overlapping features are present in both diseases. But the principal difference is that, with scleroderma, the inflammation heals with scarring and tightening of the skin. The development of diseases like scleroderma has been associated with a variety of chemicals, including polyvinylchloride, trichloroethylene, cocaine, appetite suppressant amphetamines, and adulterated cooking oils (i.e., an epidemic caused by denatured rapeseed oil—''toxic oil syndrome''—afflicted 15,000 people in Spain in the early 1980s and several hundred of them died).

Autoimmune diseases resembling lupus have been found in animals exposed to certain *metals,* including mercuric chloride, gold, and cadmium. No human reports have appeared as yet. Eosin is a chemical contained in *lipstick* that may trigger sun-sensitivity rashes and allergic dermatitis. A widely cited report speculating on a role for lipstick in the causation of lupus appeared 25 years ago, but follow-up studies have suggested that there is no connection.

## SHOULD LUPUS PATIENTS AVOID ANY FOODS OR SUPPLEMENTS?

Foods are made up of three principal components: carbohydrates, proteins, and fat. Thus, dietary manipulations can include either altering these components or raising or lowering overall caloric intake. Studies conducted on mice with lupus have suggested that high-calorie diets may accelerate mouse kidney disease, but there is no evidence that this occurs in humans. On the other hand, while ''starvation'' low-calorie or low-fat regimens help mice with lupus, they can occasionally *worsen* the disease process in humans.

Until 1989, an amino acid dietary supplement known as *L-tryptophan* was commonly taken to help induce sleep. Amino acids are the building blocks of proteins. L-tryptophan was removed from the market when it was associated with the development of a scleroderma-like disorder known as *eosinophilic myalgic syndrome,* or EMS. The 1989 EMS epidemic was traced to impurities in the manufacture of L-tryptophan, but sporadic cases of a closely related disorder called *eosinophilic fasciitis* had appeared over a 20-year period. It turns out that some cases of EMS or eosinophilic fasciitis may be related to excessive L-tryptophan ingestion in patients who metabolize the drug through an uncommon chemical pathway, which provokes an autoimmune response.

Another amino acid, *L-canavanine,* is present in all legumes but highly concentrated in alfalfa sprouts. Immunologic testing has established that this amino acid is capable of causing or aggravating autoimmune responses (see Chapter 24 for more details). I generally advise my lupus patients to avoid alfalfa sprouts but in general do not limit the intake of legumes.

A polyunsaturated fat, eicosapentaenoic acid, is a major constituent of *fish oil.* Diets enriched in this chemical ameliorate human rheumatoid arthritis and seem to help animals with lupus. But conflicting results were found when fish oil capsules were administered to humans with SLE.

Recently, one group has suggested that an autoantibody called anti-Sm (see Chapter 6), which is found in 20 to 30 percent of patients with SLE, may react with certain plant proteins in laboratory tests. This interaction theoretically suggests that this autoantibody may be able to make human lupus worse, but it has not yet been studied.

## WHAT ABOUT SUN EXPOSURE OR RADIATION?

Although correlations between diet and lupus remain ambiguous, there is a strong connection between the disease and the sun's rays. The sun emits ultraviolet radiation in three bands known as A, B, and C. The first two, ultraviolet A (UVA) and ultraviolet B (UVB), are important for lupus patients. Some scientists have suggested that when these bands of ultraviolet light hit the skin, they may damage superficial deposits of DNA, which are the body's building blocks. This results in the release of by-products that induce the formation of anti–DNA, which is known to damage body tissues.

Other mechanisms, some involving antibodies affected by sunlight known as Anti-Ro and Anti-La (see Chapter 6), might also be harmful. The binding of these antibodies to skin cells accelerates or turns on the disease process. But not all light may be bad. For example, UVA light treatments, which generally make lupus patients worse, are given to patients with psoriasis. Certain subbands of light seem to reduce inflammation. The role of sunlight and sun protection in lupus is discussed in detail in Chapter 24.

Cancer patients are frequently given radiation therapy. Despite some concerns that lupus patients could have their disease flare up when they receive radiation, this occurs very rarely and I have never seen it happen.

## WHY DO LUPUS PATIENTS FEEL WORSE WHEN THEY HAVE A COLD?

Individuals who carry the lupus "gene" can have the process turned on by a virus or bacteria. Some of the viruses implicated in SLE causation have fancy names; they include myxoviruses, reoviruses, measles, rubella, parainfluenza, mumps,

Epstein-Barr, and type C onco- or retroviruses. The evidence for this stems from the finding of elevated levels of viral antibodies in certain patients, virus particles in lupus tissue, and documentation that microbes can mimic foreign substances or antigens that turn on autoimmunity. Moreover, rats injected with certain bacterial proteins develop a rheumatoid-like arthritis. Therefore, it is easy to see how patients with autoimmune disease can experience flareups when they develop infections. Proteins made by the infectious agents may also be found in the blood or tissue of autoimmune patients and document the presence of infection.

## IS LUPUS CONTAGIOUS?

Beyond the issue of genetic transmission, there is no evidence that lupus can be spread from one individual to another. No cases of ''contagious'' lupus have ever been reported. However, several instances have been documented in which laboratory technicians handling large amounts of lupus blood samples developed positive ANAs and antibodies to lymphocytes without developing SLE. Similar claims concerning household contact between lupus patients and nonrelated individuals or the pets of lupus patients may be valid, but these reports are also not associated with the presence of disease.

## Summing Up

Our environment is full of chemicals and microbes that can induce or aggravate autoimmune diseases. In addition, sun exposure plays a role, as does diet. Only a small percentage of individuals exposed to ''lupogenic'' materials develop the disease, which suggests that certain genetic signatures greatly augment risk factors. The amount of lupogenic material necessary to bring on the disease is not known, and not everyone at risk becomes symptomatic. It is important to emphasize that if you are genetically at risk, prudent precautions to avoid unnecessary exposure are advised, but you should try not to become so fearful of the environment that you cannot function socially. While family members may share certain genetic material with a lupus patient, our research confirms that lupus cannot be spread from one person to another, as in infectious diseases. It is important to keep in mind that both genetics and the environment play a role in inducing, easing, aggravating, or accelerating lupus.

# 9

# *Drugs That May Cause Lupus or Produce Flareups*

In 1945, Byron Hoffman reported that sulfa-containing antibiotics might provoke a lupus-like syndrome. Since that time, more than seventy agents have been reported to bring on the disease. In addition, the administration of certain drugs has been found to cause a preexisting lupus to flare up. This chapter reviews the mechanisms by which this process may occur. It also looks at published data on the most problematic drugs and discusses the treatment of drug-induced lupus.

## DRUGS THAT EXACERBATE LUPUS

Most drugs that cause preexisting disease to flare up do so by acting as sun-sensitizing agents or by promoting hypersensitivity or allergy-like reactions.

### Antibiotics

> Susan was 20 years old when she began complaining of severe burning on urination. She had been diagnosed with SLE 6 months earlier and had mild disease, in the form of skin rashes and joint aches. Susan's lupus was under excellent control with Naprosyn and Plaquenil. She decided to call her gynecologist about a bladder problem and was diagnosed as having a urinary tract infection. Her physician placed her on Bactrim DS. Within 2 days, her mild cheek flush had become a bright red rash on her face, forearms, and neck. Her wrist and knees swelled up and she developed a temperature of 102°F. When she saw her rheumatologist the next day, he observed not only these physical signs but also that her white blood cell count and hemoglobin had substantially decreased. He stopped the Bactrim and put her on 40 milligrams of prednisone daily. She gradually responded to this regimen and managed to discontinue steroids within 2 weeks. She felt fine thereafter.

Like Susan, a disproportionate number of patients with lupus cannot tolerate *sulfa* derivatives. Sulfonamide-based antibiotics (e.g., sulfamethoxazole) are potent sun sensitizers. Drugs in this class include such commonly used brands as Bactrim, Septra, and Gantrisin, which are frequently prescribed to young women with urinary tract infections. Some antidiabetic drugs (e.g., Orinase), sulfasalazine (called Azulfidine, used for ulcerative colitis and rheumatoid arthritis), and diuretics also contain sulfa. Even though they are generally well tolerated, it is best to be cautious about using these drugs. *Tetracyclines* are mildly sun-sensitizing, and a greater than expected number of lupus patients are unable to tolerate *penicillin.* I tend to avoid sulfa drugs unless no alternative treatment is available but do not restrict the use of any other antibiotics.

### Nonsteroidal Anti-inflammatory Drugs (NSAIDs)

Nonsteroidal anti-inflammatory drugs (NSAIDs) are most frequently used to treat aches and pains, fevers, or pleurisy (see Chapter 26). NSAIDs run the gamut in potency from aspirin and ibuprofen to indomethacin. Though not FDA-approved for use in treating lupus, at least one of these preparations is used by over 90 percent of lupus patients at some point during their disease course.

Rarely, lupus patients taking ibuprofen (Advil, Motrin) complain of high fevers, mental confusion, and a stiff neck. This reversible noninfectious (aseptic) meningitis is practically seen only in SLE patients. Aseptic meningitis has been reported to occur with more than ten NSAIDs, but 90 percent of the published cases implicate ibuprofen. Despite this, an individual's risk for aseptic meningitis with ibuprofen is probably less than one in a thousand. To be on the safe side, I tend to use other NSAIDs in treating lupus patients.

Certain NSAIDs sensitize users to sunlight, and toxic reactions have been reported in some cases. Extra care in prescribing piroxicam (Feldene), benoxaprofen (Oraflex), and carprofen (Rimadyl) is advised. The latter two are not available in the United States. Although no longer available in the United States, phenylbutazone has been reported to cause a hypersensitivity reaction in lupus patients, which may lead to severe flareups.

### Hormones

Birth control pills in young women and estrogen replacement therapies after menopause as well as other hormonal preparations are frequently prescribed to lupus patients. Their use has been controversial, and this complex topic is covered in Chapter 17.

## DRUG-INDUCED LUPUS ERYTHEMATOSUS (DILE)

Approximately 15,000 to 20,000 new cases of prescription drug–induced lupus are reported annually in the United States. First identified shortly after the introduction of hydralazine (Apresoline) for hypertension in 1951, DILE is usually a benign, self-limited process. Although more than seventy agents have been im-

**Table 7.** *Drugs Implicated in Provoking Lupus Erythematosus*

| | |
|---|---|
| 1. Drugs proven to induce clinical lupus in at least 1 out of 1000 users | |
| Hydralazine (Apresoline) | Methyldopa (Aldomet) |
| Procainamide (Pronestyl) | D-penicillamine |
| 2. Drugs proven to induce clinical lupus in at least 1 out of 10,000 users | |
| Isoniazid (INH) | Phenothiazines (including Thorazine) |
| Sulfasalazine (Azulfidine) | Quinidine |
| Carbamazepine (Tegretol) | Griseofulvin (Fulvicin) |
| 3. Drugs rarely associated with positive ANAs and very rare clinical lupus | |
| Anticonvulsants (phenytoin, trimethadione, primidone, ethosuximide) | |
| Lithium carbonate | Captopril (Capoten) |
| Antithyroid preparation (propylthiouracil, methimazole) | |
| Beta-blockers (practolol, acebutolol, atenolol, labetalol, pindolol, timolol eyedrops) | |
| Lipid-lowering medicines (Mevacor, Pravachol, Lopid) | |
| Prazosin (Minipress) | |
| 4. Drugs that can exacerbate lupus or increase the risk of allergic reactions but do not cause lupus | |
| Antibiotics (sulfa, tetracycline—rarely, penicillins or ciprofloxacins) | |
| Nonsteroidal anti-inflammatory agents (e.g., ibuprofen) | |
| Oral contraceptives and other hormones | |
| Sulfa diuretics and diabetic drugs (Dyazide, Aldactone) | |
| Cimetidine (Tagament), alpha-interferon, and gold salts | |
| 5. Case reports have appeared implicating nearly fifty other drugs | |

plicated as inciters of drug-induced autoimmunity, 90 percent of all cases are associated with three products whose use has been decreasing: hydralazine, procainamide (Pronestyl), and methyldopa (Aldomet). If isoniazid (INH), chlorpromazine (thorazine), and D-penicillamine are added to the list, 99 percent of all clinically relevant cases can be accounted for. A more complete list of these drugs is given in Table 7.

## Epidemiology of Drug-Induced Lupus

Three principal features distinguish DILE from SLE. First, unlike SLE, DILE affects the same number of men as it does women. Second, DILE is very rare among African Americans in the United States. Also, the average age of onset for DILE is 60, as opposed to the 20-to-40 age group typical for SLE.

## A Note of Caution

If you are a lupus patient, how do you know if you have DILE? If you are prescribed a medication by your doctor, and after several weeks to months, start noticing a rash, fevers, pain on taking a deep breath, or swollen joints, consult your doctor immediately. Most DILE patients do not fulfill the criteria for systemic lupus. All seventy or so drugs implicated in DILE induce the formation of antinuclear antibody to varying degrees, but the process is self-limited. In other words, once the drug is stopped, the formation of antinuclear antibody stops as well.

Only a small percentage of these ANA-positive individuals ever develop clinical lupus. *A positive ANA does not constitute grounds to discontinue treatment with a useful drug.* Since DILE is completely reversible, the risks of not taking lifesaving heart or seizure medications, for example, are much more ominous. Moreover, there is no evidence that a lupus-causing drug administered to a patient who already has the disease will make the condition worse.

## How Do Drugs Cause Lupus?

An exciting research challenge is presented by DILE, since investigators might be able to use it as a model for understanding how lupus develops. Unfortunately, however, DILE may come about through different chemical processes unrelated to the evolution of the disease as it unfolds in most cases. Let's look at some proposed mechanisms by which drugs induce lupus.

For example, the drug can bind to a part of the cell that alters DNA. This altered DNA sets in motion an immunologic reaction, causing the body to make anti–DNA. Similarly, drugs can activate T or B lymphocytes and, as part of the immunologic response, lead to the formation of antibodies to white blood cells, or antilymphocyte antibodies—a common feature of lupus. Also, the drug may make your body so sun-sensitive that, if you are genetically predisposed, it can turn on a lupus reaction. Finally, certain drugs are broken down into chemicals, or by-products that promote the formation of autoantibodies—the antibodies that attack the body's own tissue.

Several genetic factors may also increase the risk of developing DILE. For example, the HLA-DR4 genetic marker is associated with hydralazine and D-penicillamine-induced lupus. If your liver cannot clear these breakdown products of drugs quickly, this slow clearance system is termed "slow acetylatation." If you are a *slow acetylator* and are prescribed procainamide, hydralazine, or isoniazid and if you have a certain genetic makeup, there is an increased chance that you will develop DILE. Finally, the absence of certain HLA-derived complement genes also correlates with DILE.

## Clinical and Laboratory Features of Drug-Induced Lupus

*Hydralazine* is a drug used to lower blood pressure by dilating blood vessels. When it was introduced in the early 1950s, much higher doses were used than are currently prescribed. Positive ANA tests seem to be related to high doses of hydralazine, as do longer treatment durations. For example, at 5 years, the risk of developing DILE at doses of 50 milligrams daily is zero, but it is 5 percent at 100 milligrams daily and 10 percent at 200 milligrams daily, even though at this point more than half of patients will have a positive ANA.

The average dose of hydralazine is about 50 to 100 milligrams a day. You should suspect hydralazine-induced lupus if you begin noticing joint swelling, fevers, and weight loss. Rashes or anemia are found 25 percent of the time, muscle aches and pain on taking a deep breath are less common, and organ-threatening disease is very rare. Nearly everybody with hydralazine-induced lupus has a positive ANA and specific antibodies called antihistone antibodies.

*Procainamide* is an extremely effective, often lifesaving drug that treats irregular heart rhythms. As with hydralazine, the risk of procainamide-induced lupus also depends on dose and duration of treatment. Up to 83 percent of all individuals given long-term procainamide develop a positive ANA, but only a small percentage ever develop full-blown DILE. Its symptoms and laboratory results are similar to those of hydralazine-induced lupus, except that pleurisy and pericarditis are more common with procainamide and rashes almost never occur. Organ-threatening disease is extremely rare.

Three other commonly used types of heart medicines are also culprits. *Methyldopa* (Aldomet) is an antihypertensive drug that has induced lupus, although rarely. Patients with this syndrome often exhibit a marked anemia and antibodies to red blood cells. Another drug used in treating an irregular heart rhythm, *quinidine* (Quinaglute), occasionally causes positive lupus blood tests and a severe arthritis. A group of blood pressure pills know as *beta-blockers* infrequently cause a positive ANA test, but only a handful of clinical lupus cases have been reported in connection with those drugs available in the United States (e.g., Sectral, Tenormin, Inderal, Toprol).

*Anticonvulsants* prevent epileptic seizures and may result in positive ANA tests. Nearly all clinical cases of lupus attributed to phenytoin (Dilantin) occurred in children. Occasional reports have been associated with carbamazepine (Tegretol), but phenobarbital has not been connected with DILE. Up to 15 percent of patients with lupus experience epileptic seizures. None of these drugs will cause flareups of lupus and there is no reason not to take any of them.

*Isoniazid* (INH) is an antituberculosis agent similar in structure to hydralazine. Its long-term administration may result in positive ANA tests in up to 25 percent of those taking the drug, but only a few cases of clinical lupus have been published. Drugs that treat psychosis in the *phenothiazine* family (e.g., Thorazine, Stelazine, Mellaril) may convert patients to ANA positivity up to 26 percent of the time, but lupus-related symptoms are extremely unusual. Certain preparations used to treat *hyperthyroidism* (e.g., propylthiouracil, or PTU) only in rare instances cause a lupus-like syndrome.

*D-penicillamine* is used to manage rheumatoid arthritis and scleroderma. About 1 percent of the time, these patients may develop myasthenia gravis or

SLE from the drug. Since those taking D-penicillamine already have an autoimmune disease, sorting things out can be problematic. Your doctor should consider consulting a rheumatologist. Was the diagnosis of scleroderma or rheumatoid arthritis correct all along, or could the diagnosis have been lupus? D-penicillamine has no place in the treatment of lupus and should be stopped in anyone suspected of having SLE.

## How Can We Tell DILE From SLE?

While taking a lupus-causing drug, the DILE patient displays many of the signs and symptoms seen in the patient with lupus. However, DILE patients rarely have symptoms involving the many organ systems of the body. (In other words, the central nervous system, heart, lung, and kidneys are not usually involved.) In these patients, we do find antihistone antibodies (in fact, they are found on blood testing in more than 95 percent of patients with DILE; the problem is that 40 percent with SLE also develop these antibodies). The patient with DILE does not have the other typical lupus antibodies reviewed in Chapter 6. Normal complement levels are also present in DILE patients. Further, upon discontinuing the drug, DILE improves or resolves within days to weeks in these patients. Even though antihistone antibodies decrease, the ANA test may remain positive for years.

Your doctor must differentiate DILE from bacterial or viral infections, polymyalgia rheumatica, SLE, rheumatoid arthritis, or Dressler's syndrome (fever with pericarditis in patients who have had a recent heart attack). Blood tests and cultures usually distinguish among these diseases.

## How Do We Treat DILE?

If the offending drug is withdrawn as soon as symptoms present themselves, no therapy may be necessary. Approximately one-third of the time, my patients have benefited from several weeks to months of aspirin or other NSAID medications (such as ibuprofen, naproxen, or indomethacin). If serious complications develop (e.g., pericardial tamponade or kidney disease) or the symptoms are severe (e.g., disabling arthritis, pleurisy with shortness of breath), I usually prescribe several weeks to months of moderate-dose steroids (less than 40 milligrams of prednisone daily).

Fewer that 5 percent of patients with DILE have a complicated or unfavorable course. These circumstances arise when the inciting drug is not withdrawn despite several months of symptoms or the lupus-inducing drug is reintroduced after being stopped. If you are a lupus patient and you have had fevers, rashes, or joint aches that you or your doctor feel may be from taking a lupus-inducing drug, stop it and *never* take the drug again.

# Part IV
# WHERE AND HOW CAN THE BODY BE AFFECTED BY LUPUS?

When rheumatologists evaluate an individual who might have lupus, they take a complete history, perform a physical examination, obtain appropriate blood tests, and order studies indicated by the patient's symptoms and signs. Once this information is compiled, diagnostic possibilities other than lupus must be considered and ruled out. The fourteen chapters in this part take the reader through this diagnostic process using an approach that considers symptoms, signs, and conditions affecting each organ, otherwise known as the *organ-system approach*. When the full evaluation is completed, the treating physician is able to formulate a comprehensive treatment plan.

# 10

# *History, Symptoms, and Signs*

## THE RHEUMATOLOGY CONSULTATION

A thorough medical evaluation is essential in order to make a diagnosis of lupus. This consultation should be performed by a rheumatologist or qualified internist. A rheumatologist is an internist (as are cardiologists, gastroenterologists, etc.) who has special expertise in diagnosing and managing diseases of the musculoskeletal and immune systems. The consultation includes several components. First, a history of the patient's complaints is taken. On the first visit to a consultant, it is helpful to present a summary of important symptoms and signs in a concise fashion, either by making a list of them or reviewing them mentally beforehand.

Especially when a patient has only one visit with a qualified consultant in order to evaluate the possibility of lupus, proper preparation is essential. Copies of outside records and previous tests or workups are also helpful. Lupus is not easy to diagnose; surveys have shown that the typical lupus patient consults three to five physicians before a correct diagnosis is made. In fact, studies have suggested that an average of 2 to 3 years elapse from the onset of symptoms until the time lupus is diagnosed. This interval may be as short as a few weeks to months in children, who usually display more obvious symptoms, but in patients over 60, it can be up to 4 years before a firm diagnosis is arrived at.

## THE HISTORY AND REVIEW OF SYSTEMS

The lupus consultation consists of a history, physical examination, diagnostic laboratory tests, or radiographic evaluations (x-rays, scans, etc.). A thorough initial interview is essential if the physician is to make a correct diagnosis and recommend proper treatment. After all the observations and tests are in, the doctor will discuss the findings, perhaps at the time of the visit, at a telephone conference after the initial meeting, or in a follow-up visit.

My physician interview begins by asking why the patient has come and how he or she feels. Once the patient's symptoms and history are heard, I conduct a "review of systems." As many as a hundred questions can be asked as part of this screening process. Positive responses may lead to an additional set of queries that clarify symptoms in a given area, such as how long the complaint has been present, what makes it better or worse, how it has been diagnostically evaluated and treated in the past, and its current status.

The patient will be asked about allergies, and his or her and family members' history of rheumatic disease or other diseases. Other relevant facts include possible occupational exposures to allergic or toxic substances, educational attainment, and with whom the patient lives. Unusual childhood diseases will also be explored, as well as alcohol and tobacco use, previous hospitalizations and surgeries, past and current prescription or over-the-counter medications that are frequently used. Not only do certain environmental and family histories predispose people to lupus, but this line of questioning forms the basis of a psychosocial profile that may be important in developing a productive doctor-patient relationship.

The rheumatic review of systems covers these eleven categories:

1. *Constitutional symptoms,* such as fevers, malaise, or weight loss, are dealt with first. They refer to the patient's overall state and how he or she feels. This is followed by an organ system review that goes from head to toe.
2. The *head and neck* review includes inquiries about cataracts, glaucoma, dry eyes, dry mouth, eye pain, double vision, loss of vision, iritis, conjunctivitis, ringing in the ears, loss of hearing, frequent ear infections, frequent nosebleeds, smell abnormalities, frequent sinus infections, sores in the nose or mouth, dental problems, or fullness in the neck.
3. The *cardiopulmonary* area is covered next. I ask about asthma, bronchitis, emphysema, tuberculosis, pleurisy (pain on taking a deep breath), shortness of breath, pneumonia, high blood pressure, chest pains, rheumatic fever, heart murmur, heart attack, palpitations or irregular heartbeats, and the use of cardiac or hypertension medications.
4. The *gastrointestinal* system review includes an effort to find any evidence of swallowing difficulties, severe nausea or vomiting, diarrhea, constipation, unusual eating habits, hepatitis, ulcers, gallstones, blood in stool or vomit, diverticulitis, colitis, or pancreatitis.
5. The *genitourinary* area must be approached in a respectful, sensitive way. Aside from inquiring about frequent bladder infections, kidney stones, or blood or protein in the urine, I also review any history of venereal diseases (including false-positive syphilis tests) and, in women, the obstetric history, with special attention to miscarriages, breast disorders and surgeries (cosmetic and otherwise), and menstrual problems.
6. Next, *hematologic and immune* factors that the patient may be aware of include how easily he or she bruises, anemia, low white blood cell or platelet counts, swollen glands, or frequent infections.
7. A *neuropsychiatric* history takes into account headaches, seizures, numbness or tingling, fainting, psychiatric or antidepressant interventions, sub-

stance abuse, difficulty sleeping, and—most important—what is called "cognitive dysfunction," a subtle sense of difficulty in thinking or articulating clearly.

8. *Musculoskeletal* features involve a history of joint pain, stiffness, or swelling, gout, muscle pains, or weakness.
9. The *endocrine* system review will include questions about thyroid disease, diabetes, or high cholesterol levels.
10. The *vascular* history may uncover prior episodes of phlebitis, clots, strokes, or Raynaud's (fingers turning different colors in cold weather).
11. Finally, the *skin* is discussed. The skin is a major target organ in lupus, and evidence of sun sensitivity, hair loss, mouth sores, the "butterfly rash," psoriasis, or other rashes is carefully reviewed.

In concluding the history taking, I always ask a patient whether there is anything I should know that was not covered. Ed Dubois, my mentor, dedicated his lupus textbook to "the patients, from whom we have learned." Physicians become better doctors when they listen to what patients have to say about things that the doctor may not have brought up. Occasionally, the patient hits on something in casual conversation that turns out to be quite important in shedding light on her disease.

## PHYSICAL EXAMINATION

The history and review of systems elucidate what physicins call *symptoms;* a physical examination reveals *signs*. Four methods known as inspection (looking at an area), palpation (feeling an area), percussion (gentle knocking against a surface such as the lung or liver to detect fullness), and auscultation (listening with a stethoscope to the heart, chest, carotid artery, etc.) are employed during the physical. You will be evaluated from head to toe.

First, *vital signs* are checked to ascertain weight, pulse, respirations, blood pressure, and temperature. The *head and neck* exam includes evaluation of the pupils' response to light, eye movements, cataracts, and the vessels of the eye. The ear exam searches for obstruction and inflammation. The oral cavity is screened for sores, poor dental hygiene, and dryness. I palpate the thyroid and the glands of the neck and also listen to the neck for abnormal murmurs or sounds (carotid artery bruits). The *chest* examination consists of inspection (e.g., for postural abnormalities), palpation for chest wall tenderness, percussion to evaluate fluid in the lungs, and auscultation (e.g., to rule out asthma and pneumonia). The *heart* is checked for murmurs, clicks, or irregular beats. The *abdomen* is inspected for obesity, distension, or scars; palpated for pain or hernia; percussed to assess the size of the liver and spleen; and auscultated to rule out any obstruction or vascular sounds. This is followed by an *extremity* evaluation, which

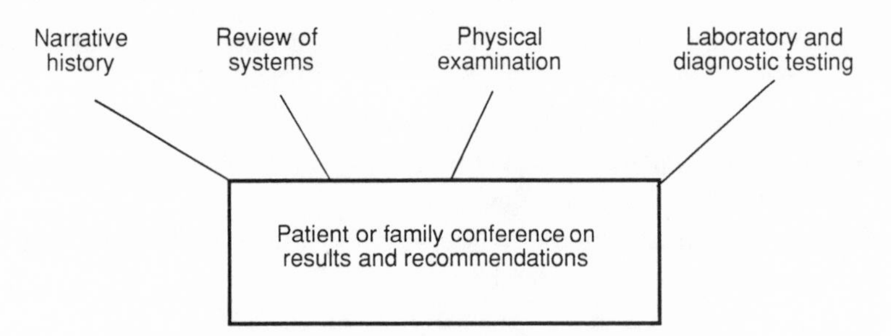


**Fig. 5.** *The Rheumatology Consultation*

includes looking for swelling, color changes, inflamed joints, and deformities. Specific maneuvers allow me to assess range of motion, muscle strength, pulses, reflexes, and muscle tone. If indicated, a *genitourinary* evaluation is done. It includes a breast examination, rectal evaluation, pelvic examination, and—in women—pap smear. In rheumatology, a genitourinary exam is necessary only if the patient has breast implants, complains of vaginal ulcers, or has other symptoms relevant to these areas. A *neurobehavioral* assessment that will reflect change in the nervous system is usually conducted as part of the ongoing conversation; neurologic deficits can be detected by observing the patient for tremor, gait abnormalities, or abnormal movements or reflexes. If necessary, a more formal mental status examination may be conducted. Finally, the *skin* is examined for rashes, pigment changes, tattoos, hair loss, Raynaud's, and skin breakdown or ulcerations.

The physical examination may include other steps as well, depending upon the problems reported and the nature of the consultation. A thorough physical examination conducted after a detailed interview allows me to order the appropriate laboratory tests. Figure 5 summarizes the rheumatology consultation.

## THE CHIEF COMPLAINT AND CONSTITUTIONAL SYMPTOMS IN LUPUS

The most common initial complaint in early lupus is joint pain or swelling (in 50 percent of patients), followed by skin rashes (20 percent) and malaise or fatigue (10 percent). Certain constitutional symptoms and signs are not included in the following chapters since they do not fall into any specific organ system; they are therefore reviewed here. These generalized body complaints consist of fevers, weight loss, and fatigue.

### Fever

Any inflammatory process is commonly associated with an elevated temperature. The official definition of a fever is 99.6°F or greater (most of us have 98.6°F as a normal temperature). Nevertheless, many patients have normal temperatures that are in the range of 96° to 97°, and what is normal for some individuals may be a fever for others. Lupus surveys published in the 1950s documented low-grade fevers in 90 percent of patients. This number has decreased to 40 percent in recent reviews—a consequence of the widespread availability and use of nonsteroidal anti-inflammatory drugs (NSAIDs), especially those over-the-counter medications that can reduce fevers. These include aspirin, naproxen, and ibuprofen (e.g., Advil, Aleve); acetaminophen (e.g., Tylenol) can also decrease fever. Many lupus patients chronically run temperatures one to two degrees above normal without any symptoms. *The presence of a temperature above 99.6°F without obvious cause suggests an infectious or inflammatory process.*

At times, I may find a normal temperature reading in a patient with a history of fevers and wonder whether or not a fever is really present. It may be helpful for certain patients to keep a fever logbook or to record temperature readings three times a day at the same hours for a week or two and to show it to their physician. Fever curve patterns may suggest different disease processes.

A low-grade fever is not usually dangerous, other than its ability to cause the pulse to rise, which decreases stamina. In certain circumstances, fevers act as warning signs of infection and suggest the need for cultures, specific testing, or antibiotics. If the temperature rises above 104°F, precautions should be taken to prevent seizures or dehydration. This might include alternating between aspirin and ibuprofen or acetaminophen every 2 hours, sponging down the head and body to lower the fever, taking plenty of fluids, or admitting the patient to the hospital. *The presence of a significant fever in any patient taking steroids* (*e.g., prednisone*) *or chemotherapy drugs should be taken very seriously* and is often a reason for hospitalization. Steroids usually suppress fevers and can mask infections.

## Anorexia, Weight Loss, and Weight Gain

Half of all patients with lupus have a loss of appetite (anorexia), with resulting weight loss. The loss of more than 10 percent of body weight over a 3-month period is rare and indicates a serious condition. Usually noted in the early stages of lupus, anorexia and weight loss are associated with disease activity. Evidence of active lupus often results in the administration of corticosteroids (e.g., prednisone), with subsequent weight gain. If your doctor has detected large amounts of protein in your urine, you have what is termed "nephrotic syndrome." Seen in 15 percent of those with lupus, this also results in weight gain.

## Malaise and Fatigue

Malaise is the sense of not feeling right. It conveys a message of aching, loss of stamina, and the "blahs." Loss of stamina or decreased endurance associated with a tendency to be tired is known as fatigue. Malaise and fatigue are observed in 80 percent of SLE patients at some point during the course of their disease; in half of these, it can be disabling. Evidence of active disease or inflammation as well as infection, depression, anemia, hormonal problems, and stress may also be associated with malaise and fatigue. The management of malaise and fatigue is discussed in more detail in Chapter 25.

## Summing Up

Before physicians can treat lupus, they must take a detailed history, perform a physical examination, and order proper laboratory tests. An accurate diagnosis encompassing all symptoms and physical signs is the goal of a consultation. The patient may have more than one process going on at the same time, and these frequently benefit from different approaches. Both patient and doctor should be flexible in how they understand the concept of disease and allow themselves to consider differing viewpoints as to what might be responsible for a specific medical complaint. Many things may contribute to fatigue, for example. And as we've seen all along, lupus is a complex disease that can be difficult to diagnose.

# 11
# *Must We Draw Blood?*

Unfortunately, the history and physical examination are not enough to diagnose or manage lupus, so yes, we must draw blood. Blood testing can make a difference. It is essential in order to diagnose a disease correctly and to gauge the patient's response to therapy. And specific tests are employed to monitor the safety of medications that might be used to treat the disease. Finally, some lupus patients have specific complications that can be diagnosed only by blood testing. This chapter offers an overview of the tests doctors order and why.

## WHAT IS CONSIDERED "ROUTINE" LAB WORK?

When a patient arrives at an internist's or family practitioner's office for a general medical evaluation, this usually calls for what doctors refer to as "screening laboratory tests." In other words, by obtaining a blood count, urine test, and blood chemistry panel, abnormalities can be detected in 90 percent of individuals with serious medical problems. This is the starting point of the workup. All the studies listed below are inexpensive and mostly automated; they can be performed within a matter of hours and do not require special expertise. Most large medical offices are equipped to perform these tests on the premises.

A *complete blood count* (CBC) is the most commonly performed laboratory test in the United States. It analyzes red cells, white cells, and platelets. Patients with systemic lupus can have low red blood cell counts or be anemic as a result of chronic disease, bleeding, from lupus medication, autoimmune hemolytic anemia (breakdown of red blood cells due to antibodies), or a vitamin deficiency. The white blood cell count can be high due to steroid therapy or inflammation or low from active lupus or a virus. Platelets are decreased when antibodies attack platelets or when the bone marrow is not making enough of them. Most patients with active SLE have an abnormal CBC.

*Blood chemistry* panels consist of anywhere from 7 to 25 tests that evaluate a variety of parameters, including blood sugar, kidney function (BUN, creatinine), liver function (AST, ALT, bilirubin, alkaline phosphatase, GGT), electrolytes (sodium, potassium, chloride, bicarbonate, calcium, phosphorus, magnesium), lipids (cholesterol, triglycerides, HDL, LDL), proteins, (albumin, total protein), thyroid function (T3, T4, TSH), and gout (uric acid). Occasionally, chemistry panels include additional studies (amylase for pancreatic function, LDH for hemolysis, iron levels, etc.), which are also inexpensive and can be done by request. Any of these tests may be abnormal in SLE patients; the reader is referred to the index to find specific discussions that address their questions.

*Urinalysis* is a useful screen for kidney involvement and urinary tract infections. All but 1 percent of lupus patients with clinically significant renal disease will have an abnormal urinalysis.

Several additional blood tests relevant to SLE also are inexpensive, readily available, and commonly ordered by rheumatologists. These consist of the *creatine phosphokinase* (*CPK*), which screens for muscle inflammation; a *Westergren sedimentation rate* or *C-reactive protein* (*CRP*), which quantitates levels of inflammation; and the *prothrombin time* (*PT*) or *partial thromboplastin time* (*PTT*), which are clotting tests and may be prolonged in those who have the lupus anticoagulant.

## ROUTINE ANTIBODY PANELS AND SCREENS

Very few doctors' offices are equipped to do reliable antibody screening. Most community and hospital laboratories are capable of performing these tests, but the lack of a national standard, inexperienced technicians, and failure to double and triple check results impair their accuracy. Autoantibodies can be tested for anywhere, but if the results are positive, they should be confirmed in a reliable rheumatology laboratory. Such facilities include university-based teaching medical centers, private facilities with a national reputation for special expertise in rheumatology (e.g., Specialty Labs, Rheumatology Diagnostic Laboratory), or certain large national lab networks (e.g., Mayo Clinic Laboratories, Scripps Clinic Laboratories). The reader is referred to Chapter 6 for a detailed discussion of all the tests listed below.

What does autoantibody screening consist of and how does your doctor interpret these results? Most of the centers listed above have what is called a ''reflex panel.'' Simply stated, a doctor orders an ANA, and if it is positive, another eight to ten antibody and immune determinations are automatically obtained.

The *antinuclear antibody* (*ANA*) is the cornerstone of rheumatic disease screening. As mentioned earlier, some ten million Americans have a positive ANA but fewer than one million have SLE. Patients with other rheumatic diseases such as rheumatoid arthritis and scleroderma as well as healthy relatives of patients with autoimmine diseases can have a positive ANA. The ANA can also become positive with aging, certain viral infections, or as a result of taking specific prescription drugs. However, less than 3 percent of patients with SLE are ANA-negative.

Practically speaking, interpretation of the ANA test is guided by a few principles. First, high levels of this antibody (greater than 1:1280 or greater than 30 International Units) are usually associated with a real rheumatic disease. Second, whereas rimed patterns are specific for lupus, homogeneous ANA patterns generally correlate with SLE, while speckled patterns are seen in SLE and with other autoimmune processes. Even though ANA levels decrease with clinical improvement, this correlation is weak at best and not a reliable gauge of the disease process. Also, different laboratories use varying standards for ANA. Therefore, a reading of 1:640 from one laboratory may be the same as 1:320 from another.

*Anti-double-stranded DNA* antibodies are rarely present in patients who do not have SLE. They are found in half of those with the disease and represent one of the more specific parameters for diagnosis and for following severe inflammation or organ involvement. Occasionally, healthy patients have low-level positive tests. If performed by a Farr or ELISA method, anti-DNA can be quantitated, and its values reflect clinical disease activity; the values are higher with flares, and they decrease with improvement.

Serum *complement* measures levels of a protein that is consumed during the inflammatory process. Along with anti-DNA, complement components C3, C4, or CH50 are the most reliable parameters for following serious disease activity. Low complement levels imply active inflammation. A few patients with genetic complement deficiencies always have low complements, and in them these determinations are not useful in following disease activity.

Levels of *anti-Ro* (*SSA*) or *anti-La* (*SSB*) are of no value in following disease activity. These tests are either positive or negative. Positive tests confirm the presence of an autoimmune problem that may not even be symptomatic and suggest that the doctor look for Sjögren's syndrome and subacute cutaneous lupus rashes while also asking the patient about severe sun-sensitivity. Young women with these antibodies should be warned about the possibility of having a baby with neonatal lupus or congenital heart block (Chapters 22 and 30). Similarly, levels of *anti-Sm* or *anti-RNP* are of little value in following patients. These tests are also either positive or negative. Anti-Sm is of no importance in following disease activity but is extremely specific and useful in confirming the diagnosis of SLE. Anti-RNP in high levels suggests that the patient may have mixed connective tissue disease (MCTD) and at low levels supports the diagnosis of SLE. The levels rarely change.

Since one-third of patients with lupus have antiphospholipid antibodies that may lead to blood clots, miscarriages, and strokes, most rheumatologists screen for these antibodies. As discussed in Chapter 21, the antiphospholipid antibodies can be tricky to detect, and a given patient may have only one of several possible autoantibodies. The most commonly ordered screen consists of an *RPR* or *VDRL* (syphilis test), *anticardiolipin antibody,* and the *lupus anticoagulant.* Further testing is necessary only if the patient's history includes a blood clot and these initial screenings are negative.

Two other tests that are frequently done are screening for the rheumatoid factor and serum protein electrophoresis. The *rheumatoid factor* test is positive in 80 percent of rheumatoid arthritis patients and 20 to 30 percent of those with SLE. High levels of rheumatoid factor with low levels of ANA suggest that a diagnosis of rheumatoid arthritis should be considered as opposed to SLE. *Serum protein electrophoresis* is an inexpensive test for blood protein abnormalities; if a broad band of gamma globulin is found, it confirms an autoimmune process.

## ADDITIONAL ANTIBODY SCREENS AND TESTS

Occasionally, additional blood testing is indicated to sort out peculiar or unusual symptoms. These tests are expensive, take longer to perform, and should be sent only to laboratories that have special immunologic expertise.

*Antihistone antibodies* are found in nearly all patients with drug-induced lupus but are not specific, since at least half of all lupus patients have them. *Antineuronal antibodies* in the spinal fluid are usually specific for central nervous system lupus. In the blood, these antibodies are positive in 20 percent of all patients with lupus, but high levels are found in 70 percent with nervous system activity. *Antiribosomal P* antibody may be positive in central nervous system disease and may correlate with SLE psychosis. *Coombs' antibody* testing screens for autoimmune hemolytic anemia, a serious blood complication of SLE.

The workup of antiphospholipid antibodies in patients with unusual manifestations of the syndrome includes obtaining *protein C, protein S, antithrombin III, platelet antibody tests,* and *noncardiolipin antiphospholipid antibodies.* Sludging of the blood—which leads to dizziness, difficulty concentrating, and a sense of fullness—is evaluated by testing for *cryoglobulin* and *viscosity,* which, if present, imply that the blood is too thick. Systemic vasculitis atypical for lupus can be assessed with *anti–neutrophilic cytoplasmic antibody* (*ANCA*) testing. Similarly, lupus-like diseases such as scleroderma or autoimmune myositis are characterized by the presence of numerous autoantibodies rarely seen in SLE, such as *anti–PM-1, anti-Jo, anti–Scl-70,* and *anticentromere antibodies.*

# 12

# *Reactions of the Skin: Rashes and Discoid Lupus*

The skin is often an affected area in lupus, with 60 to 70 percent of lupus patients reporting some skin complaint. And there is a wide range of such complaints. This chapter discusses how the skin is damaged in lupus and what it looks like under the microscope. It reviews the classifications of skin disorders in lupus as well as their principal dermatologic features. A complete discussion of remedies used for skin disorders in lupus can be found in Part V, but specific interventions that are appropriate under special circumstances are mentioned here.

## HOW IS THE SKIN DAMAGED IN LUPUS?

The sun emits ultraviolet radiation in three bands known as A, B, and C. Only the first two, ultraviolet A (UVA) and ultraviolet B (UVB), are directly harmful in lupus and are probably harmful to most of us. When these bands of ultraviolet light hit the skin of lupus patients, they damage deposits of DNA near the surface of the skin. In some mouse lupus models, radiation to the top (epidermal) layer of the skin has resulted in the generation of denatured or altered DNA, which researchers have found can lead to the formation of anti-DNA and result in tissue damage. Also, ultraviolet light can induce the production of other antibodies—anti-Ro (SSA), anti-La (SSB), and anti-RNP—in a skin cell known as a keratinocyte. Most patients who test positive for the anti-Ro (SSA) antibody are very sun-sensitive.

## CLASSIFICATION OF CUTANEOUS LUPUS

*Cutaneous* (which means relating to the skin) lupus can be broken down into three general categories:

1. *Acute cutaneous lupus erythematosus.* Almost all of these patients have active systemic lupus with skin inflammation.
2. *Subacute cutaneous lupus erythematosus* (*SCLE*). This is a nonscarring rash that can coexist with both discoid and systemic lupus but may be a "bridge" between discoid lupus and SLE.

3. *Chronic cutaneous lupus erythematosus,* also known as *discoid lupus erythematosus* (*DLE*). About 15 percent of all lupus patients are classified as having DLE, but patients with SCLE or SLE may also have discoid lesions.

The cutaneous features of lupus include *mucosal ulcerations*—sores in the mouth, nose, or vagina; *alopecia*—hair loss; *malar rash*—the butterfly rash on the cheeks; *discoid lesions*—thick, scarring, plaquelike rashes; *pigment changes*—both loss of pigment and more pigment in different places; *urticaria*—hives or welts; and *cutaneous vascular* features involving the blood vessels. Included in this category are *Raynaud's phenomenon*—when the fingertips turn red, white, and blue in reaction to cold temperatures; *livedo reticularis*—a red mottling or lacelike appearance under the skin; *purpura*—which appears like a bruise or black-and-blue mark; *cutaneous vasculitis*—breakdown of the skin due to inflammation of the superficial vessels, which can lead to *ulcers* or *gangrene,* a breakdown of the skin due to inflammation of the deep vessels or a blood clot.

There are other conditions, not usually seen in lupus patients, that involve the skin and require attention. One condition is *panniculitis*—which involves inflammation of the dermis of the skin. *Bullous lupus,* another condition, produces fluid-filled blisters or a rash similar to that of chickenpox. Complications from the use of steroids as a treatment for lupus can also induce skin damage such as *ecchymoses,* or black-and-blue marks, as well as *skin atrophy,* which results in paper-thin skin.

## DISCOID LUPUS ERYTHEMATOSUS

*Chronic cutaneous* (*discoid*) *lupus erythematosus* (*DLE*) is commonly known as discoid lupus erythematosus. It is diagnosed when a patient with a discoid lupus rash (confirmed by skin biopsy) does not fulfull the American College of Rheumatology criteria for systemic lupus (see Chapter 2); 10 percent of all lupus patients have DLE. Remember, discoid lesions may be a feature of SLE.

In the United States, 70 percent of patients with DLE are women and 75 percent are Caucasian, with the mean age of onset in the thirties. Discoid lesions appear on sun-exposed surfaces but, in rare cases, can also be found on non-sun-exposed areas. Such lesions generally do not itch. They appear as thick and scaly; under the microscope, one sees plugging of hair follicles, thickening of the epidermis, atrophy or thinning of the dermis (the part of the skin under the epidermis, which is the topmost layer of skin). Signs of inflammation are also present.

Aching joints and other constitutional symptoms are found in 10 to 20 percent of patients with DLE. Blood testing shows a positive ANA test in about half of the cases; other autoantibodies are seen in less than 10 percent. Anemia may be observed in 20 percent of the patients, and a low white blood cell count in half.

Discoid lupus can appear similar to other skin lesions. For example, rosacea, fungal infections, sarcoidosis, seborrhea, dermatomyositis, and a sun-sensitive rash called polymorphous light eruption can be ruled out by a simple skin biopsy and blood tests before diagnosing DLE.

Without treatment, discoid lesions may progress. After many years, some may turn into skin cancer. "Localized DLE" is a term coined at the Mayo Clinic in the 1930s to describe discoid lesions appearing only above the neck. They rarely evolve into systemic lupus and are treated with antimalarial drugs or local remedies. "Generalized DLE" implies lesions above and below the neck. This form has a 10 percent chance of developing into systemic lupus. Discoid lupus is managed by avoiding the sun and by using sunscreens, antimalarial drugs, and steroid creams. Sometimes steroid injections are helpful with these lesions. In rare cases, severe resistant lesions may require antileprosy drugs, oral corticosteroids, azathioprine, or nitrogen mustard ointment. See Part V of this book for a review of these treatments.

## SUBACUTE CUTANEOUS LUPUS ERYTHEMATOSUS

Subacute cutaneous lupus erythematosus (SCLE) is a rash seen in about 9 percent of lupus patients; 20 percent with SCLE also have lesions typical of discoid lupus. Unlike DLE, SCLE does not scar the skin. Under the microscope, the inflammation is mild and diffuse, not thick and scaly. The lesions in SCLE, like those in systemic lupus, also do not usually itch. The rash may look similar to that of psoriasis.

Among SCLE patients, 70 percent are women and 85 percent are Caucasian. The mean age of onset is in the early forties. Half of these patients fulfill the American College of Rheumatology criteria for systemic lupus. Among the SCLE cases, 75 percent are sun-sensitive, 65 percent have joint aches, but less than 10 percent develop organ complications (in which the heart, lungs, kidneys, or liver are involved).

We know little about the causes of SCLE, but certain drugs such as thiazide diuretics (e.g., Dyazide) have been known to bring it on. Two-thirds of SCLE patients have positive ANA and anti-Ro (SSA) tests. Nearly all have the HLA-DR3 marker, and many have a deficiency of the second and fourth components of serum complement (C2 and C4).

The lesions of SCLE are notoriously resistant to the usual drug therapies. Skin creams and antimalarials provide only modest results. But retinoid (Vitamin A) derivatives have been helpful and are important in the management of acute SCLE.

## CUTANEOUS FEATURES OF LUPUS

### Mouth or Nose Sores (Mucosal Ulcerations)

Mouth sores are seen in 20 percent of patients with systemic or discoid lupus. Occasionally, nose ulcerations are noted, which may result in nasal septal perforation. In rare cases, women may develop recurrent vaginal sores. Oral ulcers must be differentiated from herpes lesions or cold sores seen in lupus patients, especially those who are on steroids or are receiving chemotherapy. The sores may be solitary or appear as crops of lesions. They may be found on the tongue or any part of the mouth.

Oral ulcers look like lupus under the microscope. They are managed conservatively with old-fashioned remedies such as buttermilk gargles or hydrogen peroxide diluted in a few ounces of water, gargled and spit out several times a day. A local steroid, triamcinolone, which can be found in a dental gel (e.g., Kenalog in Orabase), can be applied to the lesions and usually brings about prompt healing. Antimalarial drugs and systemic steroids are also helpful. Nasal ulcers sometimes respond to a petroleum jelly such as Vaseline.

### Hair Loss (Alopecia)

There are many reasons why lupus may lead to hair loss. First of all, active disease is associated with the plugging of hair follicles, which results in clumps of hair simply falling out after being combed or washed (called "lupus hair"). Patients with discoid lupus can experience mild, generalized hair loss, bald spots (alopecia areata), or even total baldness. Steroids may induce hair loss in the male pattern of baldness—in the temples and on top of the head. Also, infections, chemotherapy, emotional stress, and hormonal imbalances are associated with hair loss. All told, about 30 percent of patients with SLE and DLE report significant hair loss.

The treatment of alopecia depends on its cause. For example, discoid lesions respond to local scalp injections with steroid preparations. If these areas form thick scars, hair may not regrow. Tapering off steroid use eliminates the "balding" pattern. Antimalarials and corticosteroids promote hair growth. Minoxidil (Rogaine) solution is a blood pressure preparation that promotes hair growth in balding men. It promotes hair growth in lupus patients but does not decrease hair loss.

### Butterfly (Malar) Rash

The term "lupus" was derived from the Latin word for "wolf" in an effort to describe one of the disease's most recognizable features. About 35 percent of patients with systemic or discoid lupus report a butterfly rash on their cheeks that suggests a wolf-like appearance. The rash reflects the angle at which the ultraviolet radiation from the sun hits the skin. Rosacea, a sun-sensitive rash called polymorphous light eruption, and other disorders can be associated with butterfly rashes, and many patients referred to me with a malar rash and a suspected diagnosis of lupus turn out to have one of these conditions and not lupus. In fact, rosacea (which is far more common than lupus) is treated with increased ultraviolet light exposure. Again, lesions in lupus generally do not itch and are identified as lupus under the microscope. One differentiating trick is that the nasolabial folds (where the outer parts of the nose and cheek meet the upper lip) are usually not affected in lupus. Occasionally, patients with a malar rash ignore their doctor's advice and apply a great deal of fluorinated (e.g., Kenalog, Synalar, Diprolene, Temovate) steroid salve for weeks or months. This results in thinning and wasting of the skin, which makes the condition look like a severe malar rash. The reason for this are that capillaries (blood vessels) near the skin become more visible and mislead the patient into believing that the lupus has worsened. Malar rashes from lupus are treated with the judicious use of steroid ointments or gels, sun avoidance, and management of lupus activity in other parts of the body.

## Changes in Pigmentation

Increased or decreased pigmentation of the skin is present in 10 percent of those with DLE or SLE. In other words, these patients may have areas of skin that are darker or lighter than expected. As inflammation heals, patients may note increases or decreases in pigment. Also, steroid deficiency (a decrease in the steroids made by the adrenal gland) can increase pigmentation, as may antimalarial drugs. *Vitiligo* is an autoimmune skin condition associated with depigmentation; it may be more common in those who have lupus. No systemic drug is helpful for the pigmentation abnormalities of lupus, but some dermatologic preparations applied to affected skin areas can improve the patient's appearance.

## Hives or Welts (Urticaria)

At some point in the course of their disease, 10 percent of patients with systemic lupus will develop hives. This is one of the few skin rashes of lupus that itch. Most cases are related to coincidental allergic reactions, but an uncommon form of lupus may be associated with "lupus urticaria." Most of these patients have deficiencies in certain blood complement components. For unclear reasons, lupus patients have an increased incidence of allergies in general (see Chapter 29). Lupus urticaria is managed with antihistamines—$H_1$ blockers such as hydroxyzine (Atarax) and $H_2$ blockers such as cimetidine (Tagamet)—antiserotonin drugs (Periactin), and steroids.

## Vascular Rashes

*Vasculitis* refers to inflammation of blood vessels, and such inflammation can lead to features detectable in the skin. Lupus usually involves the medium and small-sized blood vessels. Small arteries and capillaries under the skin can be deprived of oxygen because of inflammation, abnormal vascular tone (which controls whether a vessel dilates or constricts), or blood clots. A variety of lesions associated with lupus stem from vascular problems.

## Raynaud's Phenomenon

One-third of patients with systemic lupus exhibit an unusual sign, referred to as *Raynaud's phenomenon,* in which their fingers turn a patriotic red, white, and blue in response to stress, cold, or vibratory stimuli (e.g., jackhammers, pneumatic drills). This usually reflects a malfunction of the autonomic nervous system's ability to regulate the tone of the small blood vessels of the hand (whether they dilate or constrict). At times, Raynaud's may be observed in the feet, the tip of the nose, or on the outsides of the ears. There are many other causes of Raynaud's, and it can be seen in most other rheumatic autoimmune diseases. In fact, several surveys have shown that only about 9 percent of all Raynaud's is found among lupus patients. Raynaud's can exist by itself (as *Raynaud's disease*), but 30 to 40 percent of these patients have developed an autoimmune disease over a 10-year observation period, and some cases evolve into lupus. Infrequently, Raynaud's becomes so severe that the skin ulcerates from lack of oxygen and gangrene may develop. Raynaud's activity is usually independent of lupus activity. (In other words, active SLE does not necessarily appear at the same time as active Raynaud's, and vice versa.)

Raynaud's is managed with preventive measures. These include wearing gloves or mittens, avoiding cold environments, not smoking, and using hand warmers. Medication such as beta-blockers, decongestants, and ergots given for migraine headaches should be avoided or used sparingly. A variety of medications that increase the flow of blood to the hands may be prescribed, including nitroglycerine ointment, nifedipine (Procardia), nicardipine (Cardene), and reserpine. Severe ulcerations of the fingers or gangrene can be treated with intravenous preparations such as prostaglandin $E_1$. As a last-ditch effort to save a finger, sympathetic blocks or sympathectomies (cutting the autonomic nerves to the hand) are occasionally performed, but the results are often only temporary. Ulcers limited to one finger can be differentiated from a lupus anticoagulant-derived clot (treated with blood thinners; see Chapter 21), or cholesterol clots called emboli (treated by lowering cholesterol levels). Blood testing and studies of the blood vessels are used to determine the exact nature of such ulcers.

## Livedo Reticularis

Some 20 to 30 percent of all lupus patients have a red mottling or lacelike appearance, or livedo reticularis, under the skin. Causing no symptoms, it indicates a disordered flow in blood vessels near the skin due to dysregulation of the autonomic nervous system. Even though normal individuals may demonstrate livedo reticularis and no treatment is ever required, recent work has associated this condition with a secondary fibromyalgia, with the circulating lupus anticoagulant, or with anticardiolipin antibody. Livedo reticularis on rare occasions can lead to *livedoid vasculitis,* a condition that involves superficial skin breakage. This complication may be treated with steroids or colchicine.

## Cutaneous Vasculitis, Ulcers, and Gangrene

Inflammation of the superficial blood vessels (those near the skin) is known as *cutaneous vasculitis.* Seen in up to 70 percent of patients with lupus during the course of their disease, this finding is a reminder that more aggressive management may be required in some individuals. These lesions often appear as red or black dots or hard spots and are frequently painful. If untreated, cutaneous vasculitis can result in ulceration, or breakdown of the skin. This may lead to gangrene, or dead, black skin. Gangrene from systemic vasculitis can be a limb- and life-threatening emergency. Such a condition is usually the result of a clot from the circulating lupus anticoagulant or vasculitis of a deep middle-sized artery; a prompt workup is vital to determine whether steroids, blood-thinning agents, or both are warranted. Cutaneous vasculitis is treated less urgently with colchicine, medicines that dilate blood vessels, and corticosteroids. As with Raynaud's, sometimes prostaglandin $E_1$ infusions are used. Vasculitic lesions are not limited to the fingertips or ends of the toes; systemic vasculitis may produce ulcers on the trunk of the body or on the extremities.

## Black-and-Blue Marks (Purpura and Ecchymoses)

Black-and-blue marks that appear as blotches under the skin may result from abnormal blood coagulation in patients with active lupus. *Purpura* is the term for this phenomenon. If the purpura is small and "palpable" (that is, feels markedly different from normal skin when touched without looking), this may be a warning sign of active systemic vasculitis (see the previous section) or low platelet counts (called *thrombocytopenic purpura*). On the other hand, black-and-blue marks that cannot be felt may result from the use of nonsteroidal anti-inflammatory agents (e.g., aspirin, naproxen), or corticosteroids. Nonsteroidals can prolong our bleeding times; corticosteroids promote thinning and atrophy of the skin, which can lead to the rupture of superficial skin capillaries (called ecchymoses).

This condition is not serious and is managed by reassuring the patient that it does not represent a systemic disease process.

## OTHER SKIN DISORDERS IN LUPUS PATIENTS

Unusual forms of cutaneous lupus may occur that are not necessarily associated with any other aspect of lupus activity. Most are extremely rare, and only two are briefly mentioned here.

### Lupus Panniculitis (Profundus)

One out of 200 patients with lupus develops lumps under the skin and no obvious rash. The ANA test of these patients is often negative, and the criteria for systemic lupus are not fulfilled. Under the microscope, biopsy of the skin reveals inflamed fat pads in the dermis with characteristic features of lupus. If untreated, the skin feels lumpier and atrophies. Known as *lupus panniculitis* or *profundus,* this rare disorder responds to antimalarials and steroid injections into the lumps. Some recent reports have noted an increased number of cases that may be associated with silicone breast implants (Chapter 8).

### Blisters (Bullous Lupus)

For every 500 patients with lupus, one has bullous lesions. Looking like fluid-filled blisters or blebs (large chickenpox marks), this complicated rash can be further divided into several different subsets. Resembling another skin disease called pemphigus, bullous lupus is also called *pemphigoid lupus.* A biopsy of the skin is essential, since treatment depends on which of the three types of bullous rashes are present. Bullous lupus can be quite serious and on occasion constitutes a medical emergency, since widespread oozing from the skin can lead to dehydration and even shock. Systemic corticosteroids are frequently prescribed for this condition.

## WHAT IS THE LUPUS BAND TEST?

Since 1963, skin biopsies have been improved by our ability to determine whether a rash is mediated by the activity of the immune system, which is detected by the presence of what are called *immune deposits.* At the junction of the epidermis (top of the skin) and the dermis (the layer below the skin), we can now see if immune complexes (see Chapter 6) have been deposited. Most patients with immune disorders will have these immune complexes at the dermal-epidermal junction. Lupus is one of the few diseases in which the deposits are

**Table 8.** *Results of the Lupus Band Test*

| Diagnosis | Positive Tests in Lesions | Positive Tests in Normal-Appearing Skin |
|---|---|---|
| Systemic lupus patients | 90% | 50% |
| Discoid lupus patients | 90% | 0%–25% |
| Normal patients, sun-exposed skin | NA* | 0%–20% |
| Normal patients, non-sun-exposed skin | NA | 0% |

*NA = not applicable

"confluent" (that is, once stained with a fluorescent dye, the skin of lupus patients will reveal a continuous line).

To perform a *lupus band test,* as this procedure is called, a dermatologist takes a skin biopsy from both a sun-exposed area (e.g., forearm) and an area never exposed to the sun (e.g., buttocks). These biopsies are then transported in liquid nitrogen and stained to detect specific immune reactants (IgG, IgM, IgA, complement 3 or C3, and fibrinogen) at a pathology laboratory. The biopsy is relatively painless and leaves no scar.

A confluent stain with all five reactants or proteins implies a greater than 99 percent probability of having systemic lupus; if four proteins are present, there is a 95 percent probability; three proteins, an 86 percent probability; and two proteins, a 60 percent probability provided that IgG is one of the proteins. In discoid lupus, only lesions (areas with rashes) display these proteins. In systemic lupus, most sun-exposed areas and some non-sun-exposed areas will display these proteins.

Lupus band tests are performed for two reasons. First, the test can confirm that a rash is part of an immune complex–mediated reaction, which would indicate the need for anti-inflammatory therapy. Second, the test is performed when a patient with a positive ANA test but nonspecific symptoms does not fulfill all the criteria for systemic lupus but the physician feels strongly that a diagnosis must be made one way or the other in order to initiate treatment. Table 8 summarizes the results of the lupus band test for groups of patients.

# 13
# *Why the Aches?*
# *Arthritis, Muscles, and Bone*

The musculoskeletal system is the most common area of complaint in lupus patients. This system involves different types of tissues: the joints, muscles, bone, soft tissues, and supporting structures of joints such as tendons, ligaments, or bursae. The reader may wish to refer to Figure 6 as these areas are discussed.

## JOINTS AND SOFT TISSUES

*Arthralgia* is used to describe the pain experienced in a joint; *arthritis* implies visible inflammation in a joint. Although we have over 100 joints in our body, only those joints that are lined by synovium can become involved in lupus. *Synovium* is a thin membrane consisting of several layers of loose connective tissue that line certain joint spaces, as in the knees, hands, or hips. It is not found in the spine except in the upper neck area. In active lupus, synovium grows and thickens as part of the inflammatory response. This results in the release of various chemicals that are capable of eroding bone or destroying cartilage. Inflammatory synovitis is commonly observed in rheumatoid arthritis but occurs less frequently and is less severe in systemic lupus.

Surveys have suggested that 80 to 90 percent of patients with systemic lupus complain of arthralgias; arthritis is seen in less than half of these cases. Deforming joint abnormalities characteristic of rheumatoid arthritis are observed in only 10 percent of patients with lupus. The most common symptoms of arthritis in lupus patients are stiffness and aching. Most frequently noted in the hands, wrists, and feet, the symptoms tend to worsen upon rising in the morning but improve as the day goes on. As the disease evolves, other areas (particularly the shoulders, knees, and ankles) may also become affected. Non-lupus forms of arthritis such as *osteoarthritis* and joint infections can be seen in SLE and are approached differently.

Synovium also lines the *tendons* and *bursae*. Tendons attach muscle to bone; bursae are sacs of synovial fluid between muscles, tendons, and bones that promote easier movement. These are the supporting structures of our joints and are responsible for ensuring the structural integrity of each joint. In addition, bursae contain sacs of joint fluid that act as shock absorbers, protecting us from

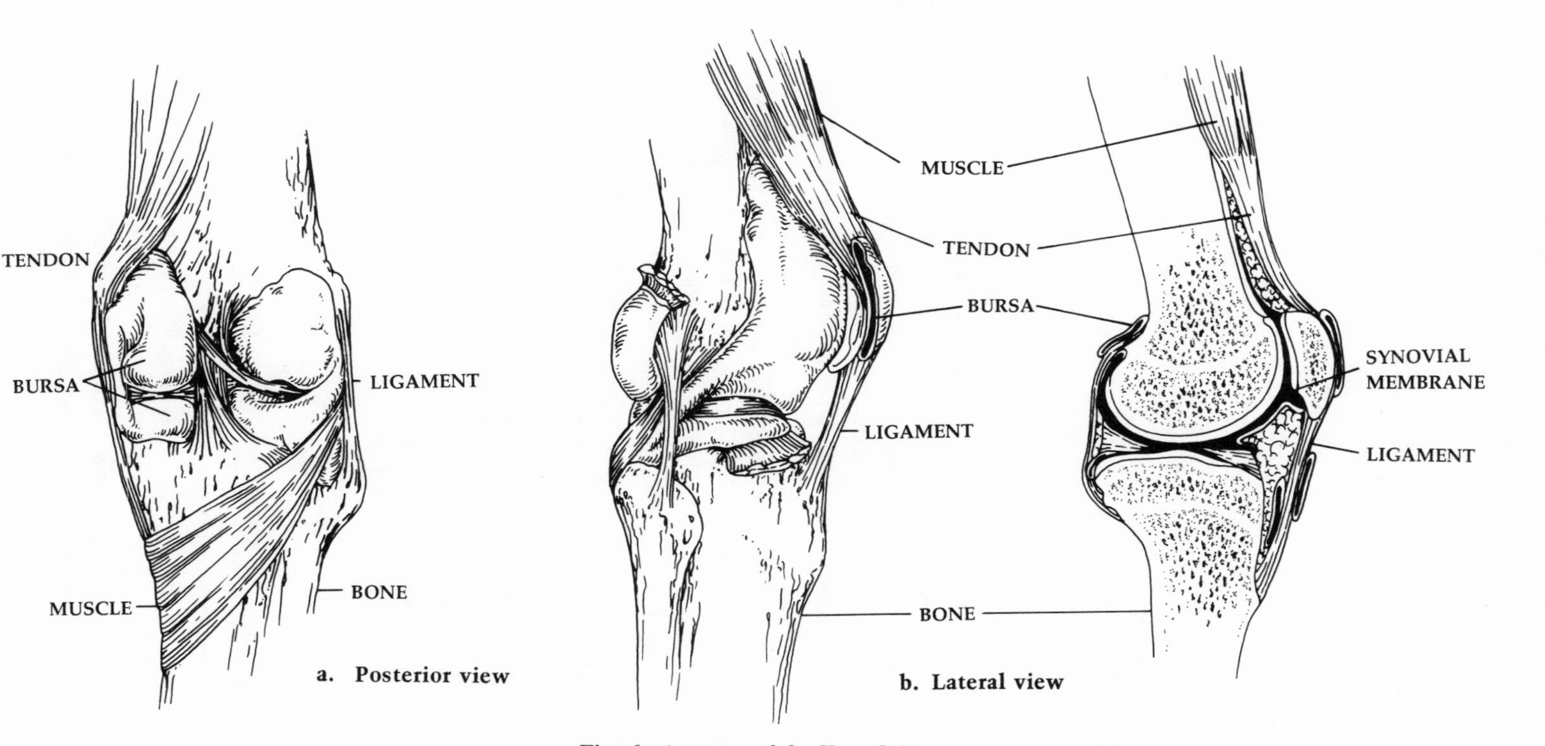


**Fig. 6.** *Anatomy of the Knee Joint*

traumatic injury. Inflammation of these structures may lead to deformity if *ligaments* (tethers that attach bone to bone) or tendons rupture. *Trigger fingers, carpal tunnel syndrome,* and *Baker's cyst* (all described in the following pages of this chapter) are examples of what can result from inflammation in these areas.

Occasionally, cysts of *synovial fluid* may form. These feel like little balls of gelatin and are called *synovial cysts.* When a joint or bursa is swollen, it is often useful to aspirate or drain fluid from the area to ensure good joint function. When viewed under the microscope, synovial fluid (fluid made by synovial tissue) provides a great deal of diagnostic information. Its analysis usually focuses on red and white blood cell counts, crystal analysis (e.g., for gout), and culture (e.g., for bacteria). White blood cell counts of 5000 to 10,000 are common in active lupus; counts above 50,000 suggest an infection; counts below 1000 suggest local trauma or osteoarthritis. Normal joint fluid contains 0 to 200 white blood cells. Crystals of uric acid (as in gout) or calcium pyrophosphate (as in pseudogout) are occasionally observed in lupus patients and necessitate specific treatment modifications. Joint fluid may also be cultured for bacteria. Viruses are rarely if ever tested for when cultures are taken (they grow poorly in cultures); but since viruses, fungi, parasites, or foreign bodies may complicate lupus, the analysis of synovial tissue obtained at synovial biopsy may also be required in addition to synovial fluid cultures.

If no evidence of infection is present when a joint is aspirated, I inject a steroid derivative with an anesthetic (e.g., Xylocaine). This often provides prompt symptomatic relief and successfully treats the swelling. Occasionally, one joint may remain swollen despite anti-inflammatory measures. In these situations, *arthroscopy* (looking at the joint with an operating microscope) along with a synovial biopsy may be diagnostically useful. A persistent *monoarticular arthritis* (arthritis in one joint only) implies either an infection, the presence of crystals, or internal damage (structural abnormalities) in the joint. The presence of *oligoarthritis* (inflammation of two to five joints) and *polyarthritis* (inflammation of more than five joints) suggests a systemic process. Joints are usually x-rayed before any diagnostic procedures are performed to make sure that no other process is being missed and also to confirm the anatomy of the area involved.

## Joints and Soft Tissues Often Involved in Lupus

Any joint lined with synovium may be affected by systemic lupus. Starting at the top, the *temporomandibular joint* (TMJ or jaw joint) can produce symptoms in up to one-third of lupus patients. Manifesting itself as jaw pain, TMJ inflammation is often confused with *myofascial pain* (a regional form of fibrositis), which consists of tense facial muscles. *Fibrositis* (see the next section) affects 20 percent of lupus patients.

The upper *cervical spine* is the only part of the spinal column lined with enough synovium to produce significant inflammation. Inflammation of this area in lupus patients is usually mild and produces pain in the back of the head. Patients who have been treated with steroids for many years can develop instability in the ligaments supporting the neck, so I may recommend that these patients use a collar when driving a car. Again, myofascial problems in the neck and upper back area are much more common than synovitis of the upper cervical spine. Myofascial pain is managed with heat, traction, cervical pillows, muscle relaxants, pain killers, and remedies that improve sleep habits.

*Shoulder* inflammation is not uncommon in lupus patients and often, when arms are raised over the head, feels the way a "bursitis" pain does. Such inflammation responds to anti-inflammatory medications or a local injection.

*Elbows* are occasionally affected in lupus patients. About 10 percent of patients may have *rheumatoid-like nodules* in this area, which feel like little peas. These nodules are much smaller than those seen in rheumatoid arthritis and are of little clinical importance except that they may cause the area below the elbow to fill up with fluid when they break down.

The *hands and wrists* are affected in most lupus patients, with swelling occurring in up to 50 percent of those with systemic lupus. Symptoms of stiffness and aching are frequently experienced in the morning hours. Since it is used more often, the dominant hand is usually more inflamed. In other words, because most people are right-handed, the knuckles of the right hand are larger than those in the left. With chronic inflammation, an "ulnar drift" or outward movement of the knuckles away from the body is observed. Inflammation of the tendons of the hand may result in *trigger fingers* (locking) or deformities in the form of contractures (shortening of the tendon) or tendon ruptures. *Carpal tunnel syndrome* has received a lot of attention in the media as a result of "repetitive stress syndrome," suffered by certain workers such as machine and computer operators. In lupus, it results from chronic swelling in the wrist, which compresses the nerves running through the wrist to the hand. Occasionally, the soft tissues show evidence of *calcinosis,* or calcium deposits under the skin. More commonly seen in scleroderma than systemic lupus, these deposits may cause occasional pain or may break through the skin, ooze, and drain.

*Costochondral margin* irritation, or *costochondritis,* produces chest pain that may mimic a heart attack. The costochondral margin is defined as the place where the sternum (breastbone) meets the ribs. Also known as *Tietze's syndrome,* costochondritis is frequently observed in lupus but is also found in many healthy young women.

Even though the lower spine is not involved in lupus, some of my patients complain of back pain, since their *hips and sacroiliac* joints are lined with synovium. About one-third complain of discomfort in this area, but destructive changes are unusual. *Knee* pain, on the other hand, is quite common in lupus patients, but since most lupus patients are young, internal damage to the knee as a consequence of athletic endeavors must also be ruled out.

The *ankle and feet* are common focuses of joint involvement. Ankles can be swollen because of fluid retention, poor circulation, or proteins that have leaked from the kidneys. In addition, swelling can occur from direct joint inflammation or altered gait as a result of foot abnormalities or deformities. Metatarsals are the foot bones in our soles that bear the brunt of our weight when we walk. In systemic lupus, the supporting ligaments of the toes loosen and produce bunions and calluses. This may evolve to the point where patients must walk on their metatarsal heads (a normally straight bone), often causing severe foot pain. If the problem is determined to be a local one, I prescribe anti-inflammatory agents and special footwear. Local injections may be used, and occasionally surgery is necessary.

## THE MUSCLES IN LUPUS PATIENTS

Two-thirds of my lupus patients complain of muscle aches, which are called *myalgias*. Most frequently located in the muscles between the elbow and neck and the knee and hip, myalgias are rarely associated with weakness. Inflammation of the muscles, or *myositis,* is observed in 15 percent of patients with systemic lupus. Established by blood elevations of the muscle enzyme CPK (creatine phosphokinase), a diagnosis of myositis necessitates steroid therapy, because muscle inflammation may cause permanent muscle weakness and atrophy if not treated. Occasionally, an electromyogram (EMG) or ''cardiogram'' of the muscles is needed to confirm the presence of an inflammatory process. Muscle biopsies are rarely necessary. In addition to physical inactivity, another cause of muscle weakness (not pain) is chronic, long-term steroid use for inflammation, which paradoxically induces muscle atrophy and wasting. In rare cases, high doses of antimalarial agents can cause muscle weakness.

A number of my lupus patients have aching muscles in the neck and upper back areas as well as tenderness in the buttocks, but these aches do not respond to steroids or anti-inflammatories. *Fibromyalgia* or *fibrositis* consists of amplified pain in what are known as ''tender points'' (in areas we have already discussed, among others). About 6 million Americans have fibrositis. Fibrositis is divided into primary and secondary types, where the primary type is from unknown causes and the secondary type is usually from viruses, trauma, or inflammation. Twenty percent of patients with systemic lupus have a secondary fibromyalgia. Fibromyalgia is associated with a sleep disorder, skin that is painful to the touch, aggravation by stress, and a relatively poor response to physical therapy. Lowering steroid doses can also temporarily aggravate fibrositis (''steroid withdrawal syndrome''). I manage fibrositis with tricyclic antidepressants—which can relax muscles, induce restful sleep, and raise pain thresholds—along with other non-anti-inflammatory interventions. It can be difficult to differentiate lupus flareups from aggravated fibrositis. (See Chapter 23 for a complete discussion of fibrositis.)

## BONES IN LUPUS PATIENTS

### Osteoporosis

*Osteoporosis,* or thinning of the bones as a result of lost calcium, may be observed in systemic lupus. It rarely causes symptoms unless it produces compression fractures and is most commonly induced by corticosteroid therapy. More than half of lupus patients require oral steroids during the course of their disease. In addition, lupus itself predisposes the bone to demineralization as a consequence of chronic inflammation and inactivity. Calcium, estrogen-containing hormones, sodium fluoride, bisphosphonates (e.g., Didronel) and calcitonin injections are used to treat osteoporosis. They are covered in more detail in Chapter 29.

### Avascular Necrosis

> Penelope went to her doctor with a sudden onset of severe pain in her right hip. She had a 5-year history of lupus, which involved her skin and joints, and had experienced recurrent bouts of pleural effusions (fluid in her lungs). Her disease was well controlled with 20 milligrams of prednisone a day. An x-ray of her hip was normal, but since Penelope rarely complained, Dr. Smith ordered an MRI (magnetic resonance imaging) scan, which revealed avascular necrosis (defined below). The pain did not respond to any anti-inflammatory medications and only an aspirin with codeine preparation provided temporary relief. Dr. Smith referred Penelope to an orthopedist who put in an artificial hip joint. She is feeling fine now.

One of the most feared consequences of steroid therapy in lupus patients is a condition known as *avascular* (or *aseptic*) *necrosis.* Experienced as a localized pain, avascular necrosis begins when fat clots produced by steroids clog up the blood supply to bone and deprive it of oxygen. This results in dead bone tissue, which, in turn, produces a tremendous amount of pain and ultimately the destruction of bone. About 10 percent of avascular necrosis is not the result of steroids but derives from clots in the blood supply to the bone in patients with the circulating lupus anticoagulant or in those with active inflammation of blood vessels (vasculitis), which obstructs blood flow to the bone. Even though avascular necrosis is seen in 5 to 10 percent of those with systemic lupus, signs of the affliction may not appear on plain x-rays for many months. Early cases may be identified by MRI. The most common target areas are the hip, shoulder, and knee. Crutches may be helpful and medications may alleviate symptoms somewhat. A limited surgical procedure known as core decompression is beneficial if performed early in selected joints, but the overwhelming majority of patients ultimately require surgery for joint replacement.

## Summing Up

Most lupus patients complain of joint aches, but only a minority demonstrate inflammatory arthritis and only 10 percent develop deformities. Inflamed synovium can cause pain and swelling in joints, tendons, and bursae. The small joints of the hands and feet are most frequently involved. Muscles aches are also present in most patients with systemic lupus, but inflammatory muscle disease (myositis) is observed in only 15 percent during the course of their disease. In order to provide optimal management, arthritis, myalgias, and myositis due to lupus must be differentiated from avascular necrosis, fibromyalgia, and the adverse effects of lupus medications.

# 14
# *Pants and Pulses: The Lungs and Heart*

Every grammar school child learns that the heart and lungs are the heart of the engine that runs the human body. If you suffer from chest pains or notice shortness of breath, you are reminded of the importance of these two organs. And for those of us treating lupus patients, studying the effect of systemic lupus on the heart and lungs is critical. Most lupus patients have complaints pertaining to the chest. This chapter will review the pulmonary and cardiac manifestations of lupus. Figures 7 and 8 depict the principal anatomic areas in the lungs and heart that are relevant to our discussion.

## THE LUNGS

When the lungs are working perfectly, they exchange oxygen for carbon dioxide effortlessly. When lupus affects the lungs, or pulmonary system, as many as eight different problems can arise that impede the ability to breathe easily. Doctors can diagnose such problems fairly accurately and quickly by using a variety of tests, including chest x-rays, pulmonary function tests, lung scans, heart ultrasounds, biopsies, drainage of lung fluid, and analysis of lung cells derived through bronchoscopy.

### The Pulmonary Workup

Surprisingly, patients may not think of mentioning their lung complaints during an interview. I make a point of asking about shortness of breath, chest pains, pain on taking a deep breath, dry cough, coughing up blood, fever, and rapid breathing. I also inquire about tobacco use and take a job history to see if there has been exposure to toxins. For example, contact with coal or asbestos can certainly damage the lungs.

In addition to a physician listening to the patient's lung, the first diagnostic tool used in evaluation is the *chest x-ray*. It is inexpensive, quick, widely available, and can detect nearly all the important syndromes discussed in this chapter. Should additional testing be necessary, a *CT* (computed tomography) *scan* or *MRI* may demonstrate structural abnormalities in the chest cavity and pleura. (The CT scan is a modified x-ray, whereas the MRI scanner uses

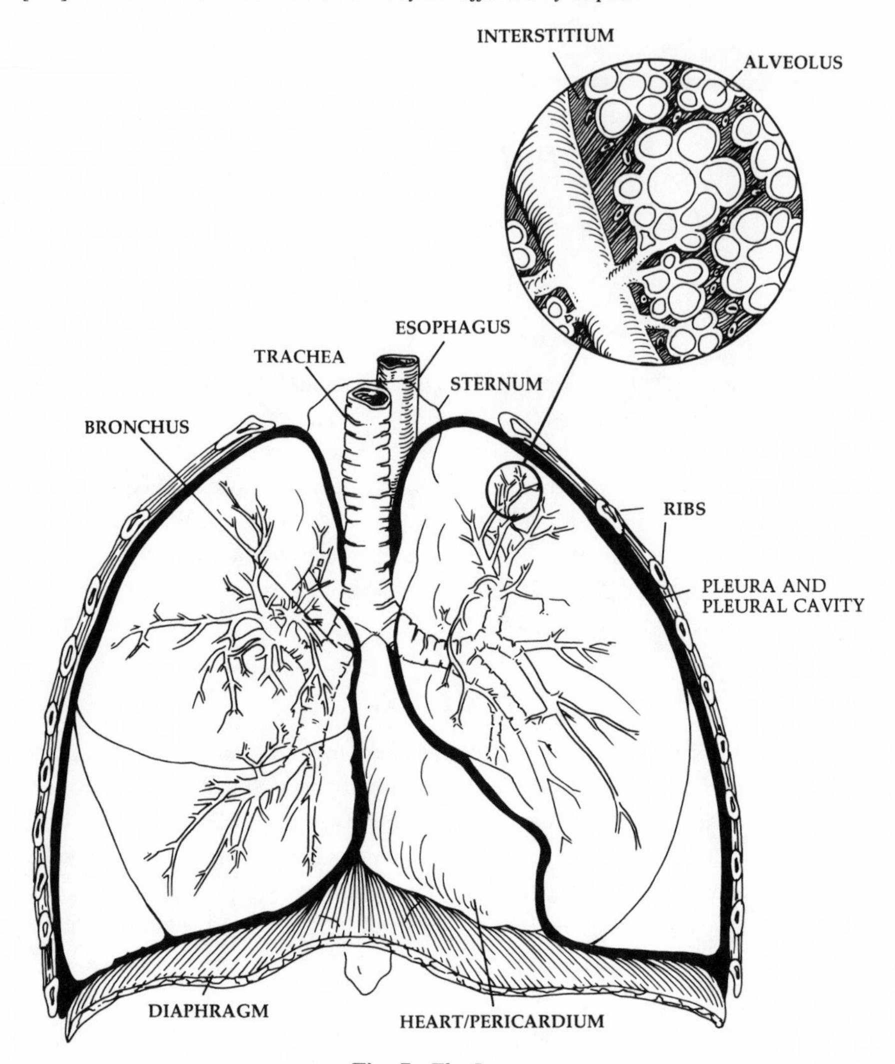


**Fig. 7.** *The Lung*

magnets to create an image and releases no radiation.) *Ultrasound* machines employ sound waves to create an image and are used to evaluate pleural disorders. Measurement of *arterial blood gases* (taking blood from the artery instead of a vein) demonstrates how much oxygen is flowing through a patient's arteries, and *pulse oximetry* is a simple, noninvasive measurement of how well the body is saturating oxygen when a patient breathes. By asking the patient to breathe in and out of a balloon (i.e., performing a *pulmonary function test*), the doctor can help to determine whether an asthmatic or bronchospastic component is present, assess lung breathing capacities, and evaluate interstitial lung function. The interstitium is tissue that provides general support to the lung and facilitates the exchange of oxygen and carbon dioxide when we breathe. Interstitial lung function is tested by the diffusing capacity, which, if low, suggests a defect in the exchange of oxygen for carbon dioxide in the lung tissue rather than in the air sac.

*Lung scans* are also used. They are painless, noninvasive nuclear medicine studies that can diagnose pulmonary artery blood clots called emboli, infection, or inflammation of the interstitium. A ventilation/perfusion scan is used to detect pulmonary emboli; an indium or gallium scan is used to detect inflammation of the interstitium. If an infection is suspected or a diagnosis of active lupus needs confirmation, doctors sometimes perform a *thoracentesis,* which is the removal of the pleural fluid for analysis. *Bronchoscopy* permits the removal of lung tissue through a flexible, thin tube for diagnostic purposes. A washing of cells obtained through this procedure is called *bronchoalveolar lavage;* it enables physicians to analyze the cell types present in the lung tissue. *Pulmonary angiograms* (whereby dye is injected into the pulmonary arteries) are the "gold standard" for diagnosing pulmonary emboli but are not without risks. A *two-dimensional* (2-D) *Doppler echocardiogram* is a simple, noninvasive ultrasound procedure that enables doctors to estimate pulmonary pressures in most patients with significant pulmonary hypertension. As a last resort for diagnosing pulmonary disease, doctors can perform a surgical procedure known as an *open-lung biopsy,* but this is rarely necessary.

## Pleurisy

John was a healthy young man until one day, upon taking a deep breath, he felt an odd sensation in his chest. Thinking it would pass, he ignored this symptom for several months. Finally, he measured his pulse and found it to be fast. John consulted his family doctor, who obtained a temperature of 99.8°F and a pulse of 120. Although he did not hear anything unusual upon listening to John's chest, a chest x-ray showed that there was fluid lining the lungs on both sides. At this point, an ultrasound estimated that John had a liter of fluid on his right side (around his lung). Since John felt well otherwise, Dr. Smith called in a pulmonary consultant, who removed the fluid. Under the microscope, the fluid showed LE (lupus) cells and an elevated white blood cell count. A blood ANA test was ordered, which came back positive, and a diagnosis of lupus was made. John was started on 20 milligrams of prednisone (a steroid) a day along with indomethacin (an anti-inflammatory drug) at 50 milligrams every 8 hours. Within 10 days, John was off steroids and holding his own with indomethacin. Plaquenil (an antimalarial drug) was then initiated, and John was able to stop indomethacin 3 months later.

### *What Is Pleurisy?*

The principal symptom of pleurisy is pain on taking a deep breath. The thin membrane sac enveloping the lung is known as the *pleura.* When it is inflamed, the term *pleuritis* is used. If fluid forms and seeps out of this membrane, there is a *pleural effusion*. The pleura does not contain lung tissue, so there should be no fear of lung disease.

Pleuritic pain (with or without effusions) is evident in 40 to 60 percent of patients with systemic lupus. Effusions are found in about 20 to 30 percent of patients, and for 2 to 3 percent of patients pleurisy is the initial manifestation of lupus. The pain can be on either side or both sides of the rib cage or it can be felt as front or back chest pain. At autopsy, more than 90 percent of lupus patients show pleural abnormalities resulting from the disease. The lining of the lungs looks similar to the lining of the heart (*pericardium*) and the lining of the abdominal cavity (*peritoneum*). Under a microscope, the pleural membranes normally consist of a few layers of loose connective tissue, but in lupus the tissue drastically thickens and shows signs of inflammation. Chemicals that the body makes as part of the inflammatory process irritate the lung and pleura and can eventually form scars and adhesions, resulting in further pain. Effusions are usually visible on a chest x-ray. Occasionally, their presence is confirmed when the patient is lying down and the fluid can be observed spreading into other body cavities. Pleural fluid moves much like water in a glass which has been turned on its side. An ultrasound or CT scan can also confirm fluid.

### *What Can Be Learned from Looking at Pleural Fluid?*

The fluid made by an irritated pleura can be clear, a *transudate,* or cloudy, an *exudate*. Transudates, associated with simultaneous irritation of the pericardium and pleura, are observed clinically when the abdomen swells (a condition known as *ascites*); when massive amounts of protein leak from the kidney (a condition called *nephrosis*); or when the kidney fails.

Exudates suggest three possibilities: active lupus in the organs, an infectious process, or a malignancy. By performing a *thoracentesis,* a procedure by which pleural fluid is withdrawn from the lung at the bedside and analyzed under the microscope, a doctor can diagnose exudative effusions. Since lupus patients are susceptible to infections, pleural fluid is usually cultured.

### *How Is Pleurisy Treated?*

In addition to infection, pleural-like pain is also observed in lupus patients who have rib fractures because of osteoporosis, trauma, or steroid use. Since 20 percent of patients with systemic lupus have fibromyalgia (see Chapter 23), the physician will have to distinguish fibromyalgia from pleurisy, which share certain symptoms (chest wall tenderness, for example).

If pleuritic pain is present but a pleural effusion is not, the doctor may try a higher dose of a nonsteroidal anti-inflammatory drug (e.g., indomethacin). Prednisone given in doses up to 40 milligrams daily is also effective but is not always necessary. Radiographically evident effusions usually necessitate corticosteroid therapy. A group of anti-inflammatory drugs known as antimalarials (e.g., Plaquenil) decrease pleural inflammation over a several-month period. Rare instances of recurrent pleural effusions may call for removal of the pleura, a procedure known as *pleurectomy* or for the introduction of irritating materials (such as talc, tetracycline, or quinacrine) into the pleural lining to prevent fluid from forming.

## Acute Lupus Pneumonitis

Over a 2-day period, Anastasia noticed a temperature of 101°F, a dry, hacking cough, and shortness of breath. She was taking 30 milligrams of prednisone and 100 milligrams of azathioprine (Imuran) a day for systemic lupus that involved autoimmune hemolytic anemia and low platelet counts. She called her family doctor, who was unable to see her and prescribed erythromycin over the telephone. Three days later, when she was still not better, Dr. Matthews obtained a chest x-ray that showed an infiltrate in the interstitial tissues. As an internist, he knew that these infiltrates are most frequently associated with mycoplasmal pneumonia; therefore he hospitalized her for intravenous antibiotics. Three days later, Anastasia was still so short of breath that she had to be transferred to the intensive care unit. A pulmonary specialist and her rheumatologist were called to see her in consultation. A bronchoscopy failed to show any evidence of infection. The rheumatologist started her on high-dose intravenous steroids, and after a few days her breathing improved.

Occasionally, lupus patients develop shortness of breath, a dry cough, pleuritic pain, and a blood-tinged sputum. More often than not, this signifies a bronchial infection or pneumonia. However, as in Anastasia's case, lupus itself can inflame the lungs and produce a condition known as *acute lupus pneumonitis* (*ALP*). Seen in 1 to 9 percent of lupus patients during the course of their disease, ALP affects the lung's interstitium (its supporting tissue), which becomes inflamed. ALP is easily observed on a chest x-ray. Most physicians are cued to look for ALP when their patients' symptoms do not clear up with antibiotics and the chest x-ray remains abnormal. When a bronchoscope is used to take a tissue biopsy of the lung, the physician will observe the interstitial areas of the lung filled with lymphocytes. Clots or vasculitis are rarely noted. Under an immunofluorescent microscope, lung tissue is stained to determine the presence of immune complexes, complement, and immunoglobulin.

If it is promptly managed with high doses of steroids, ALP can be completely reversible. Drugs that decrease steroid requirements, such as azathioprine or cyclophosphamide, are sometimes added. Despite this, it is unfortunate that up to 50 percent of patients wtih ALP die within months, often due to a delay in diagnosis.

### Diffuse Interstitial Lung Disease

Florence was diagnosed with lupus 10 years ago and had only mild aching along with sun-sensitivity. She also complained of dry, gritty eyes, and her ophthalmologist recommended that she use artificial tears. Over the next several months, Florence began having a dry, hacking cough with only occasional shortness of breath. She consulted a pulmonary specialist, who obtained a chest x-ray showing increased interstitial (within the tissue) markings. Pulmonary function tests demonstrated mild restrictive abnormalities. Dr. Hughes ordered a gallium lung scan and interstitial markings lit up, revealing where the gallium had collected. This indicated that the interstitial changes were reversible. She was started on a moderate dose of steroids and now feels a lot better.

When the symptoms described for acute lupus pneumonitis (ALP) evolve over a several-year period, it is called diffuse *interstitial lung disease* (*ILD*). This is a common complication of scleroderma, mixed connective tissue disease, and rheumatoid arthritis. It is seen in 10 to 20 percent of patients with systemic lupus. Florence's case is not atypical. Indeed, the symptoms can be so subtle that patients may not even tell their doctors about it at first. Most cases are finally identified after about 10 years of disease.

Chest x-rays along with a restrictive defect on pulmonary function testing help diagnose ILD. Lung biopsies of ILD are rarely necessary and appear similar to those of ALP, though they are more chronic and less aggressive.

If identified early, ILD is responsive to treatment with steroids. After years of the disease, the inflammation diminishes and heals with scars. At this point, treatment is not helpful, since scars do not respond to anti-inflammatory medication. Although ILD rarely leads to respiratory failure, it leads to rapid, shallow breathing with decreased stamina in its chronic phase. In order for patients to be treated properly, ILD must be differentiated from chronic aspiration (swallowed material going down the wrong pipe), adverse reactions to medication, interstitial lung infections such as pneumocystis pneumonia, and environmental irritants.

### Pulmonary Embolism

Gertrude noticed that her right leg was swollen and tender. This had never happened before, and her lupus was in remission. Her friend, a nurse's aide, told Gertrude that it might be phlebitis and that she should see her doctor. Her life was hectic and Gertrude couldn't spare the time to have it checked out. Three days later, Gertrude suddenly complained of severe chest pain on the right side and couldn't breathe. It was so severe that her sister had to drive her to the nearest emergency room. Her arterial blood gases showed low oxygen levels, so she was sent upstairs to radiology for a lung scan, which confirmed the presence of a pulmonary embolus. It turned out that a clot from the phlebitis in her leg had broken off and traveled to the lung. Her serum anticardiolipin antibodies were positive, suggesting a predisposition to form blood clots. Gertrude was admitted to the hospital, given intravenous heparin followed by oral blood thinners (Coumadin), and discharged a week later, very lucky to be alive.

One-third of lupus patients have antiphospholipid antibodies such as anticardiolipin (see Chapter 21), and one-third of these patients have a clot or thromboembolic episode during the course of their disease. One-third of these episodes consist of blood clots that travel through the blood and settle in the vessels of the lung; once there, they are called *pulmonary emboli*. Even though blood clots can be caused by something other than lupus, surveys have reported that 5 to 10 percent of patients with systemic lupus will sustain a pulmonary embolus at some time in their lives.

A patient with a pulmonary embolus will complain of acute shortness of breath and chest pains. The diagnosis is confirmed if the patient exhibits a combination of low levels of oxygen in the blood and a lung scan that shows a mismatched defect (oxygen not being exchanged for carbon dioxide in the blood vessels). In some cases, a pulmonary angiogram (dye injected into the pulmonary vessels) may be needed to verify the diagnosis. The chest x-rays may look normal at first, but after several days most show a pattern of infarction (dead tissue) with a wedge-shaped defect representing collapsed, unoxygenated lung tissue.

For lupus-mediated pulmonary emboli, treatment is threefold: A patient initially gets an intravenous anticoagulant (heparin) followed by oral medication; warfarin (Coumadin) for at least several months; and then often a daily low dose of aspirin. Some patients with very high levels of anticardiolipin antibody or those who have recurrent emboli on low-dose aspirin are treated with warfarin indefinitely.

## Pulmonary Hemorrhage

Less than 1 percent of patients with systemic lupus sustain bleeding or hemorrhage into the air sacs of their lungs (*pulmonary hemorrhage*), but up to 10 percent of the deaths from all forms of active lupus are associated with such an event. Manifested by coughing up blood, pulmonary hemorrhage usually occurs early in the course of the disease, and children are unusually susceptible. Pulmonary hemorrhage may look like ALP, but it progresses more rapidly and targets the air sacs (alveoli) of the lungs. The chest x-ray in a patient with pulmonary hemorrhage will reveal fluffy alveolar infiltrates; in addition, active, multisystem organ disease is often present. Bronchoscopy (whereby a narrow tube with a telescope is placed in the airways through the mouth) will reveal hemosiderin (blood pigment) and stained macrophages, but vasculitis is rare. Immune complexes, blood complement, and immunoglobulin can be detected through characteristic stains. How blood gets into the air sacs is unknown, but the process can be triggered by active lupus or a concurrent infection. Pulmonary hemorrhage is usually fatal, but a patient may be saved by aggressive, quick management that combines high-dose steroids and a chemotherapy known as cyclophosphamide with or without a blood-filtering treatment know as apheresis, in which antibodies and other chemicals that promote inflammation are removed.

### Pulmonary Hypertension

George was mildly short of breath and, after waiting about a year, finally consulted his family doctor. He had been diagnosed with discoid lupus 20 years earlier and had occasional aches, but otherwise he felt well and worked full time. His doctor obtained a normal chest x-ray but was concerned enough to order pulmonary function tests. These were only mildly abnormal. A pulmonary consultation was requested, which resulted in ordering a two-dimensional Doppler echocardiogram. It showed elevated, twice normal pulmonary pressures. Further blood testing for antiphospholipid antibodies was negative. By now, George was winded on minimal exertion and had to stop working. He was brought to the hospital for a trial of intravenous nifedipine (a blood vessel dilator), which failed to lower his pulmonary pressures. George is currently on the waiting list for a lung transplant, since he cannot hope that his own lungs will recover adequate function.

Along with pulmonary hemorrhage, pulmonary hypertension is one of the most feared complications of systemic lupus. Mildly increased pressure in the pulmonary arteries is observed in 10 to 15 percent of patients with lupus and is generally without symptoms, but significant rises (above 50 millimeters of mercury) can be life-threatening. Like George, most patients feel well at first; later on, they start experiencing mild shortness of breath or a sense of breathlessness. Higher pressures are often associated with a sense of breathlessness, but chest x-rays are surprisingly normal and reveal pulmonary function tests with only mildly restrictive changes. Under certain circumstances, pressures are measured by a 2-D Doppler echocardiogram (a special heart ultrasound) and more accurately but invasively by a Swan-Ganz catheter, where a pressure gauge is wedged into the pulmonary artery. If a patient experiences a rise in pressure because of pulmonary emboli, the condition may reverse itself with the administration of anticoagulant drugs. However, pulmonary hypertension (a measure greater than 50 millimeters of mercury) in patients without antiphospholipid antibodies (see Chapter 21) is almost always fatal within 5 years unless a lung transplant is performed. Drugs called vasodilators (e.g., hydralazine or nifedipine) open up blood vessels and may help temporarily; they are often given in a hospital with the aid of a Swan-Ganz or similar pressure monitor. The cause of antiphospholipid antibody–negative pulmonary hypertension in patients with systemic lupus is not known, but it probably has something to do with the lining of pulmonary arteries of the lung being sensitive to a variety of chemicals.

## The "Shrinking Lung" Syndrome

This uncommon complication of systemic lupus is manifested by a sense of breathlessness and decreased chest expansion in patients who have elevated diaphragms on chest x-rays. The diaphragm is a muscle that weakens as a result of pleural (lining of the lung) scarring or other proposed mechanisms. The shrinking lung syndrome may respond to steroid therapy, but is is usually a benign, nonprogressive process.

## Adverse Reactions to Medications Used in Lupus

A small but significant group of individuals are sensitive to aspirin or other salicylates (aspirin products) and develop wheezing (bronchospasm) when they take salicylates. Similar reactions can also occur, to a lesser extent, with other nonsteroidal anti-inflammatory drugs (e.g., ibuprofen, naproxen). Drugs used to treat autoimmune disease—such as methotrexate and gold—can induce interstitial lung disease and occasionally provoke acute asthma. Steroids and chemotherapy drugs can also make a lupus patient more susceptible to pulmonary infections. These possibilities must be ruled out before any of the lung conditions discussed above are considered.

## Summing Up

If you suffer from lupus, are short of breath or it hurts to take a deep breath, have a dry cough, cough up blood, wheeze or have chest pain, your lungs are probably involved. Inflammation of the pleura (the lining of the lungs) is common and usually not serious. Interstitial lung disease (which involves the supporting structures of the lungs) and pulmonary emboli are the second and third most frequently observed lung complications in systemic lupus. They can be managed with steroids and drugs that thin the blood and prevent clots. Acute lupus pneumonitis, pulmonary hemorrhage, and pulmonary hypertension are very serious conditions and are difficult to treat, even in the most experienced hands. Infections, malignancies, allergic reactions, and other processes must be ruled out before a diagnosis of lupus-related lung disease can be arrived at successfully; a variety of well-understood diagnostic tests are available to do this.

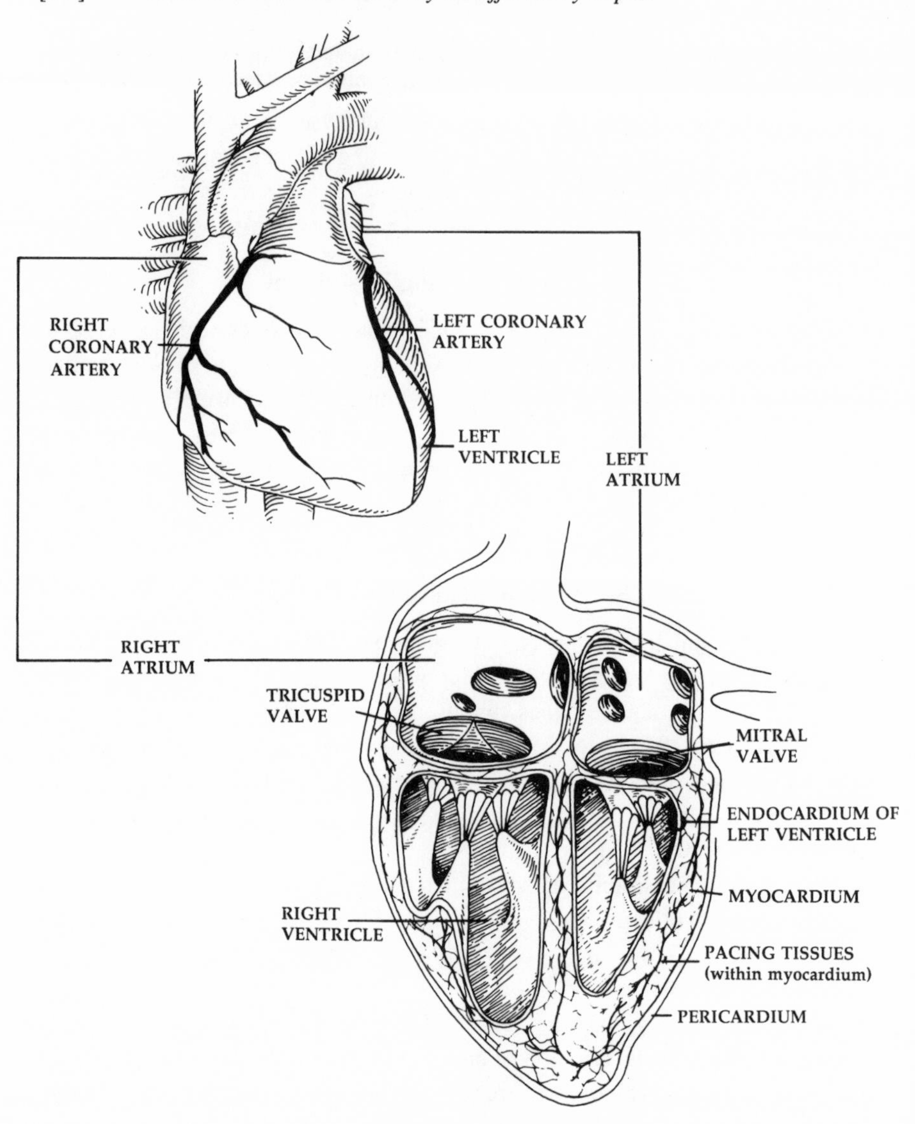


**Fig. 8.** *The Heart*

## THE HEART

Abnormalities of the heart can also significantly undermine the lupus patient's quality of life and, like lung diseases associated with lupus, heart disease can be serious. The major cardiac symptom, chest pain, is reviewed in detail here, along with heart diseases in lupus patients. Unfortunately, blood testing is not particularly useful in looking for active heart disease. Instead, there are many sophisticated tools that allow physicians to assess and manage the cardiac manifestations of lupus.

## Why Do Lupus Patients Get Chest Pains?

Esther had systemic lupus with symptoms of occasional pleurisy, a mild anemia, and swollen wrists. Her disease was adequately controlled with ibuprofen and Plaquenil. One evening, while reading in bed, she experienced the sudden onset of severe chest pressure. She told her husband, who thought it might be a heart attack, and he took her to the emergency room. Esther's chest x-ray and electrocardiogram were normal. She saw her family doctor the next day, who ordered a two-dimensional echocardiogram (a heart ultrasound) to look for evidence of a pericardial effusion (fluid around the heart sac). This turned out to be negative. Eventually, an upper gastrointestinal endoscopy examination was performed, which showed an erosive gastritis from ibuprofen with evidence of esophageal reflux (where food is pushed back up into the esophagus from the stomach). She was started on an antireflux and antiulcer regimen consisting of omeprazole (Prilosec) and told not to lie down right after eating a big meal. Ibuprofen was discontinued, and Esther was much better within a week.

Chest pain is a common but potentially misunderstood feature of systemic lupus. Fortunately, it only rarely indicates heart disease, but it can be traced to many causes, each requiring its own method of management. I find it important to exercise patience and have an open mind when I am interviewing patients with chest pain because of the complexity of its diagnosis and treatment.

The most frequent cause of chest pain in lupus is *esophageal*. Difficulty swallowing, called reflux esophagitis, and a digestive disorder involving the movement of food have been reported in half of those with the disease. Autoimmune diseases and anti-inflammatory medications used to treat them may also induce esophageal pain. (The diseases, because of altered motility and the medications, can erode or ulcerate tissue lining.) This type of pain is managed with antacids and medications that improve the movement of food and block acid release, both of which are reviewed in detail in Chapter 18.

The second most common type of chest pain is *chest wall pain*. This is easily diagnosed; the patient feels extreme discomfort when pressure is applied to the breastbone or, more technically, the costochondral margins of the sternum. It can be a symptom of fibromyalgia or a syndrome know as costochondritis (also called Tietze's syndrome). Aspirin, local moist heat, and occasional injections into tender points are the treatments of choice.

Fluid that settles around the sac of the heart, or *pericardium,* can also produce a central chest pain that occurs when the patient is at rest, which is relieved by leaning forward. *Angina pectoris,* or decreased blood flow to the coronary arteries with pain resulting from lack of oxygen, is just as likely if not more likely to occur in lupus patients as in anybody else. However, it often has an earlier onset in patients taking long-term steroids that can be a sign of coronary vasculitis or myocarditis as well as atherosclerosis of the coronary arteries (discussed later in this chapter). Chest pain of *pulmonary origin* is common with pneumonia or a pulmonary clot (embolus) and may have nothing to do with lupus.

### Other Cardiac Symptoms

A persistently rapid heartbeat, called *tachycardia,* is a feature of active lupus resulting from inflammation or fevers. Other causes of tachycardia, such as infections or an overactive thyroid gland, must be considered. Rapid heartbeats are treated with anti-inflammatory medications and occasionally beta-blockers (e.g., Inderal, Tenormin). *Irregular heartbeats* are noted when the myocardium (heart muscle) or pacing system of the heart is inflamed or scarred. *Shortness of breath* can be due to lung disease (such as asthma) or a failing heart muscle that cannot provide adequate cardiac output.

## CARDIAC DISORDERS

### What Is the Pericardium?

Heather was recently diagnosed with lupus on the basis of skin rashes, fatigue, and a positive ANA (antinuclear antibody) test. She began experiencing chest pressure, a low-grade fever, and a rapid pulse associated with shortness of breath. The chest pain diminished in intensity when she leaned forward. Her doctor obtained a chest x-ray, which showed an enlarged heart shadow as well as small pleural effusions. An electrocardiogram revealed inflammation of the pericardium, and the diagnosis of pericarditis was made. When a 2-D echocardiogram (heart ultrasound) showed a moderate amount of fluid in the pericardium, a diagnosis of lupus serositis was also made (fluid in both the pleura and pericardium) and Heather was started on 40 milligrams of prednisone daily. Shortly thereafter, indomethacin was added to her treatment. Within 3 weeks, she was able to discontinue the steroids, but she had to take indomethacin for another month. Plaquenil (an antimalarial) was also begun, and a year later she has yet to have a recurrence.

The term *pericardial effusion* is used to describe fluid around the sac of the heart. Present in 50 percent of patients with lupus who have undergone an ultrasound (the technical name is two-dimensional echocardiography), the condition is usually without symptoms and may not require any specific measures. During the course of their disease, 25 percent of my patients, like Heather, complain of chest pains below the breastbone, which are frequently relieved by bending forward. The pain correlates with abnormalities in what are called "ST segments" on an electrocardiogram (ECG). These individuals have *pericarditis,* or inflammation of the pericardium, the sac surrounding the heart. Evidence of prior pericarditis is found microscopically in 60 percent of autopsied lupus patients. In pericarditis, the pericardial fluid contains several thousand white blood cells, and areas of the sac show that lymphocytes and plasma cells are present in pericardial tissue. By listening through a stethoscope, a doctor can hear a harsh rubbing sound, or *pericardial rub,* in up to 20 percent of those with acute pericarditis. Pericarditis does not imply an organ-threatening disease, since the heart tissue is not involved. If the pericardial fluid does not show evidence of infection, acute pericarditis is managed with high-dose nonsteroidal anti-inflammatory drugs (e.g., indomethacin) along with a short course of moderate-dose corticosteroids.

On rare occasions, pericarditis is complicated by *pericardial tamponade* or *constrictive pericarditis.* Tamponade occurs when so much fluid is formed that the heart muscle is prevented from pumping blood adequately. This life-threatening complication necessitates immediate removal of the fluid, or "windowing," to allow the fluid an outlet. Chronic inflammation can promote adhesions and scars in the pericardial lining, which can result in signs similar to those of tamponade. Constrictive pericarditis, as this process is called, is cured surgically by stripping or removing the affected pericardial tissue, a procedure known as a *pericardiectomy.*

### What Is Myocarditis?

Shanna was 20 years old and had already had lupus for 8 years. Her course had been very rocky and was complicated by kidney involvement and one attack of seizures from central nervous system vasculitis. At this point, Shanna became concerned about her appearance and was reluctant to take the steroid dose her doctors recommended; she "forgot" to take it two to three times a week. One morning she began experiencing low-grade fevers, a rapid pulse, and a dull chest pressure. Her electrocardiogram showed nonspecific abnormalities with an elevated blood creatine phosphokinase (CPK—a muscle enzyme released during a heart attack or with muscle inflammation). A myocardial infarction (heart attack) was ruled out, but her continued symptoms warranted admission to the coronary care unit. A cardiologist was called to see her in consultation, and since her clinical status was unclear, he recommended a cardiac catheterization. When it was performed the next day, a myocardial biopsy was also obtained. Shanna was diagnosed with lupus myocarditis. High doses of intravenous steroids were administered immediately, and a rapid improvement in Shanna's condition was noted.

Underneath the pericardium lies the *myocardium,* which is the body's heart muscle. Each time this pump contracts, we experience a heartbeat. The blood supply to this muscle is provided by the *coronary arteries.* Ten percent of lupus patients eventually experience myocarditis, or inflammation of the heart muscle. In fact, at autopsy, 40 percent of patients with systemic lupus show evidence of prior myocardial involvement. The symptoms of myocarditis usually include a rapid pulse and chest pains and frequently coexist with active, systemic lupus. Chest x-rays frequently show an enlarged heart, and signs of congestive heart failure may be evident. Infections, particularly viruses, can also induce myocarditis. It is important to remember that lupus patients are very susceptible to a variety of infections. Creatine phosphokinase (also called CPK)—a muscle enzyme that is released with trauma, inflammation, and heart attacks—is often elevated when the blood is tested. Since it may not be possible to differentiate a heart attack from lupus myocarditis or viral myocarditis, a *heart catheterization* might be undertaken. Inflamed heart muscles can decrease blood flow through coronary arteries. Catheterization consists of injecting dye into the coronary arteries to determine how open they are and whether the heart is receiving enough oxygen. The procedure may also include an *endomyocardial biopsy,* in which a tissue sample of the myocardium is taken. Under the microscope, lupus myocarditis reveals the presence of plasma cells and lymphocytes.

Because myocarditis is a serious complication of systemic lupus, the condition is treated with high doses of corticosteroids for at least several weeks. Other measures include medications that alleviate heart failure and coronary artery insufficiency.

### Congestive Heart Failure and Myocardial Dysfunction

Raphael was 50 years old and his lupus had been in remission for over ten years. Nevertheless, when he was in his twenties and thirties, his disease had been quite active and he had taken moderate doses of steroids for quite a while. During this time, he developed high blood pressure and elevated cholesterol levels. His blood sugars were borderline high. As long as he watched his diet, limited his salt intake, and walked for a half hour twice daily, Raphael felt well. He was on a mild blood pressure medication. When his daughter graduated from college in Boston, he flew from the West Coast with his wife to attend the ceremony. Unfortunately, he forgot to take his medicine with him, and when he called his clinic on a weekend to have them call it in to a Boston pharmacy, he was unable to get an immediate response. That evening, he went to a seaside restaurant and splurged. After consuming a 3-pound lobster in drawn butter with french fries, Raphael began wheezing and complaining of shortness of breath. Within 3 hours, he couldn't breathe and had to be taken to an emergency room. His blood pressure was 200/130 and his chest x-ray showed pulmonary edema (water in the air sacs). Diuretics were administered intravenously. Within an hour, he had urinated over 800 milliliters and felt much better.

A failing heart muscle cannot pump enough blood into the arteries and tissues to maintain normal body functioning. *Congestive heart failure* results when either the left or right side of the heart fails to pump enough blood. In right-sided heart failure, fluid accumulates in the ankles, the liver enlarges, and the neck veins become fuller. A failing left heart, as in Raphael's case, pushes fluid back into the lungs, which may lead to *pulmonary edema.* In the past, 5 to 10 percent of lupus patients developed congestive heart failure, although recent surveys suggest that—as a result of general medical advances and improvements in life-style and diet—its incidence is decreasing.

Failure can be brought on or aggravated by the long-term administration of corticosteroids, anemia, hypertension, serositis, fevers, or disorders of the heart valves (see below). To minimize heart failure, patients are advised to restrict salt in their diets; or they might be prescribed diuretics, drugs that reduce pressure on the heart muscle; digitalis, a drug that makes the heart pump more efficiently; or other agents that dilate blood vessels.

Two-dimensional Doppler echocardiograms (heart ultrasounds) and studies of wall motion have shown that many patients with active lupus have subtle abnormalities in the left side of the heart that affect filling and pumping, but they show no evidence of congestive heart failure. Termed *myocardial dysfunction,* these abnormalities reflect a low-grade myocarditis or inflammation-induced stress on the heart muscle. These abnormalities usually disappear when lupus is in remission.

## Libman-Sacks Endocarditis and Other Valve Disorders

Eliza had three miscarriages before her physician found the presence of antiphospholipid antibodies upon blood testing. Her doctor diagnosed her as having a very mild case of lupus that did not require treatment. Several weeks later, after she had had several teeth repaired and root canal work done, Eliza complained of feeling lethargic. Her doctor detected a low-grade fever of which Eliza was not aware, but he also obtained a blood count that revealed anemia, and this needed attention. When the fevers persisted and became more pronounced, blood cultures were obtained that were positive for a bacterium called *streptococcus.* A diagnosis of subacute bacterial endocarditis was made when her two-dimensional echocardiogram showed a bacterial growth, known as a vegetation, on her aortic valve. Eliza was hospitalized and treated with intravenous antibiotics for 3 weeks, but she became increasingly short of breath. A cardiac catheterization was performed which suggested severe impairment of an aortic valve. She was taken to surgery, where her aortic valve was replaced. At surgery, evidence for Libman-Sacks endocarditis was found, with the complication of an infected vegetation.

The inner surfaces of the heart, particularly those that line the four heart valves, are known as the *endocardium.* Materials such as cellular debris, proliferating cells, and immune complexes that may average 1 to 4 millimeters in diameter, which may be found on the endocardium in lupus patients, are called *vegetations.* Originally described by doctors Emanuel Libman and Benjamin Sacks in 1923, *Libman-Sacks endocarditis* is almost exclusively found in patients with antiphospholipid antibodies (see Chapter 21). Clinically manifested as a heart murmur, the vegetations are usually so small that they are detectable by a two-dimensional echocardiogram only 30 percent of the time. Transesophageal echocardiograms (an ultrasound performed after a tiny tubelike camera is swallowed to rest in the esophagus) increase detectability to about 60 percent. Although they alter the dynamics of the heart only 1 to 2 percent of the time, Libman-Sacks vegetations have two potentially serious complications. First, they are prone to become infected, which leads to what is called *subacute bacterial endocarditis,* where the vegetation is a growth site for bacteria. This condition has a high mortality rate and may necessitate a cardiac valve replacement. As in Eliza's case, a vegetation can become infected after a visit to the dentist. *Hence, I advise all patients with antiphospholipid antibodies to take prophylactic antibiotics before and after any dental procedure to prevent the valve from receiving infected materials swallowed during surgery to the mouth.* Second, portions of the vegetations that flake off and travel into the brain through the carotid artery may cause a cerebral clot or stroke. *Therefore, I advise all patients with established vegetations that they should be treated prophylactically with low-dose aspirin or other drugs that reduce the ability of platelets to promote clotting or initiate anticoagulation therapy.*

Damage to the mitral, aortic, pulmonic, or tricuspid heart valves is found only slightly more often in patients with lupus than in the general population. Pulmonary hypertension in these patients is associated with damage to the tricuspid valve, which results in a condition know as tricuspid regurgitation (see Figure 8).

For unclear reasons, a floppy mitral valve, or *mitral valve prolapse,* is probably more prevalent among lupus patients. Palpitations, chest pains, and fibromyalgia are also associated with this syndrome, which is managed with beta blockers (e.g., Tenormin, Inderal), antibiotics for dental or surgical procedures, and a decrease in caffeine intake.

### Coronary Artery Disease and Myocardial Infarction (Heart Attack)

Bonnie was 38 years old and had been treated for lupus since she was 20. For the last 3 years, she had been maintained on dialysis. Prior to that, she had taken high doses of steroids along with chemotherapy for active systemic lupus. While on vacation, she developed sudden pressure in her chest and visited the local emergency room. The emergency room doctor told her she was too young to have heart disease, gave her antacids, and sent her home after obtaining a normal electrocardiogram. When the pain persisted, she returned to the hospital several hours later. This time, her electrocardiogram showed an evolving heart attack. She was admitted and eventually had a three-vessel bypass.

Atherosclerotic heart disease (hardening of the arteries) has become the third most common cause of death in lupus patients, following complications of kidney disease and infection. Because more young women are surviving other early organ-threatening complications of lupus, heart complications are now becoming more prevalent after 10 to 20 years of disease. The side effects of long-term use of moderate- to high-dose steroids include hypertension, diabetes, hyperlipidemia (e.g., high cholesterol), and ultimately premature atherosclerosis. Many healthy-appearing patients of mine in their thirties and forties, such as Bonnie, have developed angina or sustained myocardial infarctions. I now routinely place those patients at risk on low-dose aspirin, which inhibits clotting, or hydroxychloroquine (Plaquenil), which lowers cholesterol levels by 15 to 20 percent as well as inhibiting clotting. I also screen some of these patients for additional risk factors, for example, I may test them annually on a treadmill to look for coronary artery disease.

In rare circumstances, active lupus is associated with chest pains (angina pectoris) from inflammation or vasculitis of the coronary arteries, or *coronary arteritis*. A coronary angiogram shows the characteristic narrowing of arteries in this unusual condition. It is treated with corticosteroids to prevent a heart attack. Interestingly, children as young as 5 years of age have been reported to develop acute reversible coronary arteritis.

### Hypertension

Monitoring blood pressure during each office visit is a good medical practice. Hypertension, defined as a blood pressure greater than 140/90, is observed in 25 to 30 percent of patients with systemic lupus. The most common causes are kidney disease and long-term steroid use. There are no special considerations unique to managing hypertension in lupus patients, since it usually responds to conventional regimens. Untreated or inadequately treated blood pressure can minimally cause headaches, but hypertension can also lead to stroke, cardiac failure, and heart attack.

## The Electrocardiogram and Conduction Defects

The electrocardiogram (ECG) is an inexpensive, harmless, and readily available tool that provides the doctor with numerous clues in screening for heart problems. An ECG detects the heart rate, rhythm, anatomic orientation, and heart chamber size. Moreover, it can suggest whether the lining of the heart (pericardium) or heart muscle (myocardium) is inflamed and whether the coronary arteries are damaged or in danger. Although one-third of lupus patients may display abnormalities on an ECG, a normal ECG does not rule out heart disease.

An ECG can also assess the heart's pacing system, which may be abnormal in lupus patients. Ten percent of patients with systemic lupus have pacing abnormalities or electrical conduction defects that lead to palpitations or missed beats as a result of inflammation or scarring of the heart tissue. The heart's electrical or pacing mechanism travels through this damaged tissue, and these abnormalities result when electrical signals are interrupted.

Lupus present at birth (neonatal lupus) can be characterized by varying degrees of heart blockage, since autoantibodies (e.g., anti-Ro) cross the placenta and can damage fetal pacing tissues. Conduction abnormalities in the heart may be treated with drugs that control the resulting arrhythmia, but insertion of a pacemaker is occasionally required. (See Chapter 22 for a more complete discussion of neonatal lupus.)

## Summing Up

Lupus patients frequently complain of chest pain that may or may not be related to heart disease. The sources of true cardiac pain most frequently involve the lining of the heart, which rarely points to a serious disorder. Myocardial disease, which involves the heart muscle, is often serious and may include inflammation in the form of myocarditis or heart muscle dysfunction, which occasionally produces congestive heart failure. In patients with antiphospholipid antibodies or the circulating lupus anticoagulant, the inner surface of the heart is predisposed to developing vegetations. This may lead to valve damage or stroke. Coronary artery disease and high blood pressure appear prematurely in patients taking steroids for the long-term treatment of lupus. It is essential that doctors take complaints of chest pain seriously and establish its source in order to treat it properly. Finally, the pacing mechanism of the heart can be impaired by scarring from previous inflammation or by the presence of autoantibodies that preferentially settle in pacing tissue.

# 15

# *Heady Connections: The Nervous System and Behavioral Changes*

When my patients tell me they feel as though their brains had been fried, I have to know what is going on in order to help them. For many reasons, the central nervous system (CNS) and its related behavioral changes in lupus are the most misunderstood and mismanaged aspects of the disease. The numerous signs and symptoms that are found may indicate any of twelve clinical and behavioral syndromes associated with systemic lupus. Since all of these syndromes are managed differently, carefully honed diagnostic skills on the part of your doctor will be important. Working together, patients, allied health professionals, and physicians can optimize the management of nervous system lupus.

## HISTORICAL NOTES

Neurologic involvement in lupus patients was first mentioned in 1875 by Moriz Kaposi, who described altered mental status and coma in a terminal patient. Sir William Osler, often called the father of modern medicine, described several patients with CNS lupus at the turn of the century. The first modern study appeared in 1945, and most of the reports through the 1960s dealt with CNS vasculitis, or inflammation of the blood vessels in the brain. A landmark report from the Massachusetts General Hospital in 1968 reviewed brain sections at autopsy in twenty-four individuals with lupus and surprisingly concluded that vasculitis in the brain was quite rare. As a result, for nearly 20 years there was a great deal of confusion regarding the classification of CNS lupus. In the early 1980s, Nigel Harris and his colleagues described an antiphospholipid antibody known as anticardiolipin and were able to show that as many as one-third of all acute CNS events were due to clots traveling to the brain or formed within the brain and not due to active lupus. This was followed by the description of cognitive dysfunction as a unique complication of systemic lupus by Judah Denburg and his colleagues in the early 1980s. New classification systems of neurologic lupus have been proposed. We review the relevant findings here.

## WHAT ARE THE MAJOR NEUROLOGIC MANIFESTATIONS OF LUPUS?

The principal neurologic manifestations of lupus are shown in Table 9. Most are present in several of the syndromes associated with CNS lupus.

The most common symptom of CNS involvement is *cognitive dysfunction*. Characterized by confusion, fatigue, memory impairment, and difficulty in articulating thoughts, it may be present by itself (discussed in its own section below) or as a part of active lupus. Cognitive dysfunction may also appear as a component of vasculitis, lupus headache, and organic brain syndrome, among other syndromes.

*Headaches* are a feature of the ''lupus headache'' syndrome (discussed later in this chapter), but they can also be a manifestation of cognitive dysfunction, CNS vasculitis, hypertension, or fibromyalgia.

*Seizures* result from acute brain inflammation, scarring from prior vasculitis, acute strokes, or reactions to medications used to treat the disease, such as corticosteroids or high-dose antimalarials.

*Alterd consciousness*—such as stupor, excessive sleepiness, or coma—is observed with CNS vasculitis, but can be induced by medication or an infectious process.

*Aseptic meningitis* is an acute condition that involves inflammation of the lining of the spinal cord (meningitis), but spinal fluid cultures do not grow out bacteria, viruses, or fungi. This usually indicates either CNS vasculitis or a reaction to ibuprofen (Motrin, Advil).

On occasion, a patient may develop *paralysis*. The most common causes range from strokes related to lupus anticoagulant clots or paralysis due to clots induced by anti-phospholipid antibody. Vasculitis of the covering of the spinal cord, infection, or bleeding can also induce paralysis.

*Movement disorders* such as tremor, writhing motions termed *chorea,* and balance deficits (known as *ataxia*) imply disease in areas of the brain containing the basal ganglia or cerebellum. Almost any of the twelve syndromes discussed below can be responsible for movement disorders.

*Altered behavior* includes psychosis (e.g., losing touch with reality), organic brain syndrome (e.g., including a demented mental state), depression, and confusion. This behavior is differentiated from cognitive dysfunction in that these alterations are obvious to physicians and family, whereas the cognitive changes are usually subtle and often noticed only by the patient. Chemical imbalances, active disease, infection, or reaction to medication may cause altered behavior, whereas cognitive dysfunction without other symptoms usually derives from a blood flow abnormality.

When *strokes* are due to lupus, they result from high blood pressure, low

**Table 9.** *Major Neurologic Manifestations of Lupus*

| |
|---|
| Cognitive dysfunction (e.g., not thinking clearly) |
| Headache |
| Seizure |
| Altered mental alertness (e.g., stupor, coma) |
| Aseptic meningitis |
| Stroke |
| Peripheral neuropathy (e.g., numbness, tingling, burning of the hands and/or feet) |
| Movement disorders |
| Paralysis |
| Altered behavior |
| Visual changes |
| Autonomic neuropathy (e.g., flushing, mottled skin) |

platelet counts, antiphospholipid antibodies, and long-term use of steroids with premature atherosclerosis or active vasculitis.

*Visual changes* may be caused by inflammation of the optic nerve, clots resulting from antiphospholipid antibodies, medication such as steroids or antimalarials, or an uncommon condition known as pseudotumor cerebri (which involves swelling of the optic nerve).

Finally, *peripheral* or *autonomic nerves* may produce numbness, tingling, or local nerve palsies (e.g., inability to lift up the wrist).

A careful history and thorough physical examination are critical components of the diagnostic evaluation, which allows physicians to determine which syndrome they are dealing with. These syndromes are discussed below.

## LUPUS SYNDROMES OF THE NERVOUS SYSTEM

The manifestations of lupus discussed above can be part of numerous syndromes. This section examines how they look clinically and reviews the management of the twelve principal CNS syndromes due to lupus.

### Central Nervous System Vasculitis

Elyse developed a rash while sunbathing with her high school friends. When it did not go away, her mom took her to see their family doctor, who ultimately diagnosed lupus. She was slightly anemic and had a small amount of protein in her urine. Upon taking 20 milligrams of prednisone daily, her tests started to improve and her rash disappeared. Three weeks later, she caught a flu that was going around, and a few days later, began noticing difficulty concentrating and connecting her thoughts. She had a low-grade fever which quickly rose to 104°F. Dr. Baker started her on antibiotics and a decongestant. However, Elyse developed severe headaches and a stiff neck; she also started convulsing. She was admitted to the hospital. A rheumatologist and neurologist were called in to see her in consultation. Magnetic resonance imaging of the brain was normal, but her spinal tap was consistent with CNS vasculitis based on an elevated protein and cell count as well as large amounts of antibodies to nerve cells. No evidence for infection was found. Elyse was transferred to intensive care and given high doses of intravenous steroids. She worsened for several days and became practically comatose but eventually began to respond. She was discharged 3 weeks later and, after convalescing for a month at home, finally returned to school on 20 milligrams of prednisone a day.

Vasculitis of the CNS is an inflammation of the brain's blood vessels due to lupus activity. The most serious of the CNS syndromes associated with lupus, it was the first to be described and is one of only two CNS syndromes (the other is psychosis) that fulfill the American College of Rheumatology criteria for defining lupus.

Vasculitis of the CNS usually occurs early in the disease course; over 80 percent of episodes take place within 5 years of diagnosis. The typical patient experiences high fevers, seizures, meningitis-like stiffness of the neck, and may manifest psychotic or bizarre behavior. Ten percent of lupus patients develop CNS vasculitis. Untreated, their course rapidly deteriorates into stupor and ultimately coma.

***How Is CNS Vasculitis Diagnosed?***

Vasculitis of the CNS can be definitively diagnosed in several ways. The physician may order a conventional angiogram (an x-ray study of the vessels of the brain after injecting them with dye), a magnetic resonance angiogram, or tests to detect high levels of antineuronal antibodies in the serum. These tests may be negative in CNS vasculitis, but if positive, they usually correlate with evidence from blood testing for active lupus (elevated sedimentation rates, low complements, high anti-DNA, etc.).

Some of the most helpful diagnostic studies are obtained using a spinal tap. There is no need to fear this procedure. Also known as a lumbar puncture, it sounds scarier than it is. Spinal taps are usually only minimally uncomfortable. It is, however, important to lie flat in bed for at least 8 hours afterwards to avoid a post–spinal tap headache. Spinal fluid, which is obtained from a spinal tap, may indicate increased white blood cells or protein, LE cells, or low sugar levels. I usually order a test known as a MS panel, so called because it was originally used to diagnose multiple sclerosis. This panel may show data suggesting an immune reaction (e.g., elevations in IgG synthesis rates, Ig levels, or the presence of oligoclonal bands) in the central nervous system. Antineuronal antibodies are often present in large amounts.

### *How Is CNS Vasculitis Treated?*

CNS vasculitis is treated with high doses of corticosteroids, which are often given intravenously. Very high dose or pulse steroid dosing (Chapter 27) may be instituted. If improvement is not noted fairly quickly, cyclophosphamide is given intravenously with or without a blood filtering treatment know as apheresis (Chapters 27 and 28). The mortality rate of CNS lupus has decreased over the last 30 years as a result of improved diagnostic testing and greater awareness of this syndrome. In the 1950s, the majority of patients with acute episodes did not have a good chance of surviving; today, 80 percent of patients recover from such episodes. Unfortunately, doctors sometimes treat a condition that is not present, since CNS vasculitis mimics various syndromes, some of which we now look at.

## The Antiphospholipid Syndrome

Monica thought she had licked lupus. Ten years into the disease, she was off all medication, felt well, and was working full time as a designer. Glad to be off Plaquenil and the anti-inflammatory Lodine, Monica had lots of plans for the future. One day, while braiding her niece's hair, Monica suddenly found herself unable to move her right arm. She saw a doctor at the clinic she attended and was told that she had had a stroke. A brain MRI was ordered, which showed several small defects pointing to prior CNS episodes. The neurologist Monica was referred to confirmed the diagnosis of stroke and, in view of her history of lupus, started her on 60 milligrams of prednisone daily to rule out CNS vasculitis. However, the spinal tap results were negative, as were all blood tests for lupus activity. Monica then decided to see her rheumatologist in the next town, who stopped her prednisone and detected, through blood tests, the presence of antiphospholipid antibodies. Monica's right arm began to return to normal after several months and, as a long-term preventive measure, her rheumatologist placed her on lifelong low-dose aspirin.

Anyone who has antiphospholipid antibodies (Chapter 21) has an added risk of developing blood clots that can travel to and settle in the brain. Whether from a Libman-Sacks endocarditis (Chapter 14) or other sources, clots to the arteries or veins (called arterial emboli or venous thromboses) account for 10 to 35 percent of all sudden CNS complications in lupus. The disease does not have to be active; sedimentation rates, blood complements, and anti–DNA levels may all be normal.

Blood clots (called thromboembolic events) to the brain usually occur suddenly and are not associated with pain. Patients may find themselves unable to move an arm or leg; they may develop slurred speech or acute weakness in a particular part of the body. Eventually, MRI and CT show a focal defect in the affected area of the brain. Spinal fluid is usually normal. Some patients are fortunate in having warnings of these events, which may disappear after seconds, minutes, or hours and are called *transient ischemic attacks,* or *TIAs.*

I recommend that all my patients with certain types of antiphospholipid antibodies take low-dose aspirin on a continuous basis to try to prevent clots. I treat clotting episodes aggressively with anticoagulant medications or drugs that inhibit certain actions of blood platelets that are responsible for clotting blood, and I avoid using steroids unless lupus activity is present.

## Other Coagulation and Flow Abnormalities

Complications in the CNS may arise when your blood becomes too thick or too thin or when it clots too easily. Sjögren's syndrome is found in up to one-third of patients. Patients with Sjögren's complain of dry eyes, dry mouth, and arthrits. These patients have particularly high autoantibody levels on blood testing. Very large amounts of antibody, particularly IgM antibodies with or without Sjögren's, can increase the viscosity (thickness) of blood.

Several autoimmune disorders including lupus may be complicated by a condition known as *hyperviscosity syndrome,* in which blood takes on the quality of sludge and is accompanied by symptoms of confusion, dizziness, and mental clouding. Occurring as a result of an excessively large number of autoantibodies, it is treated with chemotherapy and blood filtering (apheresis). Similarly, a circulating protein that becomes solid in cold temperatures known as *cryoglobulin* is occasionally found in SLE. Because it "clogs up the works," this disorder, known as *cryoglobulinemia,* is very similar to hyperviscosity syndrome, sharing many of its symptoms. Cryoglobulinemia is more common than hyperviscosity syndrome and is usually looked for when the skin is ulcerated, has a mottled appearance, or shows vasculitic skin lesions. Cryoglobulinemia responds to chemotherapy, blood filtering, and interferon treatments.

Alterations in blood clotting, which can lead to bleeding from blood vessels in the brain and therefore can mimic strokes, are observed when platelet counts become very low. Two examples of this are *idiopathic thrombocytopenic purpura* and *thrombotic thrombocytopenic purpura.* The reader is referred to Chapter 20 for more details.

## Lupus Headache

Naomi had suffered from headaches for years, but this was the headache from hell. Her lupus was responding fairly well to treatments and seemed under control. And even though she told her parents not to worry all the time, Naomi was a chronic overachiever and her parents couldn't help but be concerned. In the past, Fiorinal or an occasional aspirin with codeine preparation had stopped Naomi's headaches. Also, she found that when she lowered her stress level and practiced biofeedback, her headaches lessened. But neither medication nor relaxation helped this time. Her doctor gave her an injection of sumatriptan (Imitrex), which gave Naomi some temporary relief. When she was in his office, he drew blood and was puzzled to find her lupus to be more active than usual. Upon seeing these results, Dr. Metzger prescribed 20 milligrams of prednisone daily for a week, and the headache finally disappeared.

At least once a day, one of my patients calls me and explains that his or her head is going to "split open." Compared with the general population, lupus patients are perhaps twice as likely to suffer from migraine-like headaches. These headaches seem to coincide with dilation of the cerebral blood vessels, but we still don't know their cause. Many patients also have antiphospholipid antibodies, while others display Raynaud's phenomenon (see Chapter 12), which, interestingly, is caused by restriction of the blood supply to the hands and feet. This instability in the tone (ability to dilate or constrict) of the blood vessels, which allows them to be easily altered, may result from a defect in local autonomic nervous system control.

The diagnosis of lupus headache involves a careful consideration and ruling out of other causes of headache, including high blood pressure, lupus medications that can cause headaches (e.g., methotrexate, omeprazole, indomethacin), brain infections, or other cerebral pathology, such as an aneurysm, tumor, or malformation of brain vessels present from birth. The physician can rule out these possibilities by taking a history, performing a physical examination, and, if necessary, ordering an MRI scan of the brain.

Lupus headache is managed much like conventional migraine in that painkillers (analgesics) such as Fiorinal, sumatriptan (Imitrex) injections, anti-inflammatories like Naprosyn, and vasoconstrictors such as Cafergot are used for acute attacks, while beta-blockers, tricyclic antidepressants, or calcium channel blockers offer a degree of prevention and may sometimes be taken all the time. Lupus headache, however, is different from most migraines, since patients may respond to a 1 week trial of 20 to 60 milligrams of prednisone daily, which is rarely useful to most migraine sufferers. Any lupus patient with a headache that does not respond to routine measures deserves a neurologic workup.

## Lupus Myelitis

This rare but serious complication of lupus can include paralysis or weakness that ranges from difficulty in moving one limb to quadriplegia. In lupus myelitis, the sac encasing the spinal cord is inflamed or blood clots are formed in the spinal arteries. Half of all lupus myelitis stems from antiphospholipid antibodies and half is from active vasculitis. The physician will probably first administer steroids to treat any possible inflammation. Anticoagulant drugs, such as heparin, are frequently added.

## Fibromyalgia

Angela was a shy woman without a great deal of self-confidence. When she was diagnosed with systemic lupus, her self-esteem plummeted to an all-time low and she had a terrible time coping. Angela read everything she could about lupus and was almost obsessive in her desire to control her disease. She consulted four rheumatologists before she found one who was responsive to her needs. She called Dr. Jones three times a day about every new development she thought he should know about. Her most troubling complaint was memory loss. Over the past few months, Angela had had difficulty remembering dates and suffered from severe headaches and muscle spasms. She read in medical textbooks, about severe CNS disorders and became convinced that she had vasculitis. On the basis of her symptoms, Dr. Jones ran a blood panel that revealed a minimally elevated sedimentation rate and a slight increase in DNA antibody. He then prescribed 40 milligrams of prednisone a day, and she felt a bit better the first week. In time, however, her skin became so sensitive that it could not be touched. Also, Angela gained 20 pounds over the next few weeks. Her sister, who worked in a medical office, insisted that Angela see a physician who had treated her earlier in the year. At this office visit, Dr. Wolfe explained the difference between CNS vasculitis and lupus with fibrositis, which he had diagnosed a year earlier. He showed how fibrositis can mimic inflammatory (or vasculitic) symptoms and explained that steroids make fibrositis worse. Angela was started on Elavil (a tricyclic antidepressant that, in low doses, promotes sleep, relaxes muscles, and decreases pain perception) for fibrositis, and she was quickly tapered off the steroids.

Many of my lupus patients may complain of difficulty sleeping and cognitive problems (e.g., decreased ability to concentrate) similar to those observed in the chronic fatigue syndrome as well as lack of stamina and chronic muscle tension headaches. Fibromyalgia, or fibrositis, is a syndrome that makes one very sensitive to pain; it afflicts 6 million Americans. Even though its cause is unknown, up to 20 percent of patients with systemic lupus have a concurrent fibromyalgia syndrome characterized by at least 11 of 18 specific tender points throughout the body and increased pain in the soft tissues. The administration of steroids, with an adjustment of their doses, induces most of the fibromyalgia I see in lupus patients, although poor coping mechanisms and untreated inflammation leading to a secondary fibrositis are also, less commonly, causes of the syndrome.

Since medications used to treat lupus do not help fibromyalgia and corticosteroids can worsen its symptoms, I make an effort to rule out active lupus before assuming that fibromyalgia is causing these symptoms. This can be tricky and difficult. Tricyclic antidepressants, muscle relaxants, and drugs that suppress the reuptake of a chemical in the brain called serotonin are used to treat fibrositis. Fibrositis is reviewed in depth in Chapter 23.

## The Peripheral Nervous System

> Sarah was an accomplished artist when she was told she had lupus. Fortunately, it was relatively mild and controlled with antimalarial drugs. One day, having recently gotten over a cold, she was sitting at her easel in a sunny spot out of doors. When she tried to stand up, however, she could not raise her left foot above her ankle. For the past few weeks, Sarah had noticed intermittent symptoms of numbness and tingling in her left leg and foot, but she thought nothing of it. Now, she was sent to a neurologist, who performed a muscle study called an electromyogram (EMG) along with a nerve conduction study. Her ''foot drop'' turned out to be consistent with mononeuritis multiplex resulting from active lupus. Sarah was started on high doses of steroids for a few weeks and made a complete recovery.

I occasionally come across patients with symptoms of numbness and tingling with or without a burning sensation or an inability to move part of the body. These problems fall into the domain of the peripheral nervous system, which consists of the twelve cranial nerves that are found in the face as well as nerves emanating from the cervical, thoracic, lumbar, and sacral spine. These nerves are divided into motor and sensory roots. Impairment of motor nerves leads to problems with movement, ranging from Bell's palsy in the face to a wrist or foot drop in the extremities. Sensory defects produce numbness and tingling. The CNS includes the spinal cord and brain and is distinguished from the peripheral nervous system for purposes of diagnosis and treatment.

Between 10 and 20 percent of patients with lupus exhibit inflammation of the peripheral nervous system at some point. Peripheral neuropathies can result from inflammation of the nerves, a consequence of lupus (which is also called *mononeuritis multiplex*), or compressed nerves, which can result from lupus synovitis, as in *carpal tunnel syndrome*. Many other conditions are associated with peripheral neuropathies and must be considered in what physicians term a ''differential diagnosis.'' These include fibromyalgia-induced numbness and tingling (which produces normal EMGs and nerve conduction studies), diabetes, kidney failure, neuropathy, or a herniated disc.

Inflammation in peripheral nerves is managed with short courses of moderate- to high-dose corticosteroids, and compressed nerves may respond to anti-inflammatories, local injections, splinting, or surgical decompression.

### The Autonomic Nervous System

When you sweat, urinate, or have palpitations, your autonomic nervous system is at work. These are the body functions we have some control of and rarely think about—like breathing and the heart rate. The autonomic nervous system is our ''flight or fright'' response to any form of stress. This system regulates adrenalin release, the tone of local blood vessels, and muscular contractions. Rapid or slow pulse rates, sweating, feelings of hot and cold, rapid or slow bladder and bowel transit times, and burning sensations are counted among our autonomic responses. Although inadequately studied in lupus, the autonomic nervous system may also be impaired in many patients.

## NERVOUS SYSTEM SYNDROMES THAT AFFECT BEHAVIOR

### Cognitive Dysfunction

Neil was an accountant with a large firm and nobody knew he had lupus. Since having been treated successfully for pleurisy and joint swelling a year before, Neil would not even think about his illness, and his general practitioner had him on no treatment other than occasional aspirin. Over the past month, Neil had encountered difficulty in remembering the names of his secretary's husband, the postman, and the parking attendant. If he tried hard enough, the names came to him, but it took a few minutes. Neil was surprised to find that none of his coworkers noticed his poor recall; it was so bad that he had to look at his appointment schedule every hour to remember what to do next. He returned to the consulting rheumatologist, who initiated a neurologic workup that uncovered nothing unusual. However, when Neil was given some psychological tests for memory and other thinking abilities, some odd results turned up. Neil denied depression or fibrositic pain. His rheumatologist placed him on Plaquenil and 3 months later added Atabrine. Although Neil showed a 70 percent improvement, he still had to take it easy and pace himself in order to function well at work.

I frequently encounter lupus patients who complain of confusion as well as profound fatigue, difficulty in articulating thoughts, and memory impairment. Blood testing may confirm evidence of systemic lupus, but other tests may be normal. Conventional spinal fluid evaluations and brain imaging frequently show no abnormalities, and these individuals often look well. A superficial examination of mental function will detect no deficiencies, leaving the physician puzzled. When this occurs, some well-meaning physicians may explain to their patient that they are depressed, stressed, or having difficulty coping. They may have lupus, but the disease could not be causing their symptoms.

### *How Can I Convince My Doctor This Really Exists?*

Even though this scenario is still all too common, studies began to appear in the early 1980s showing that lupus could be responsible for a whole host of subtle cognitive difficulties. Behavioral testing of lupus patients revealed that up to 70 percent at times had decreased ability to focus, deficits in attention span and task completion, altered memory, and decreased problem-solving capabilities. Only 20 percent of control subjects had similar difficulties.

The cause of cognitive dysfunction is not known, but it is probably mediated by two factors. First, circulating chemicals such as cytokines (Chapter 5) may induce the syndrome. Some of these cytokines work differently in lupus patients, particularly interleukin-2 and the interferons, and have been shown to cause cognitive dysfunction when they are administered to patients with advanced cancer or chronic hepatitis. However, the symptoms of cognitive dysfunction are often intermittent.

Recent work has suggested that blood-flow abnormalitics (termed hypoperfusion) may play a role. For example, on a newer form of brain imaging known as PET (positron emission tomography) scanning, lupus patients with cognitive dysfunction display areas of the brain that do not receive enough oxygen, and this correlates with their symptoms. This $5000 test is usually not medically necessary in diagnosing the syndrome, since it can be detected by routine neuropsychological testing (paper-and-pencil tests) or a cruder form of PET called SPECT (single-photon-emission computed tomography), which costs only $1000.

Cognitive dysfunction must be differentiated from depression, fibromyalgia, behavioral alterations due to medication, infections, strokes, or other brain disorders. In certain patients with cognitive symptoms, we sometimes find increased antibody levels that react to nerve cells in the spinal fluid and evidence of active lupus on blood testing.

### *Can Anything Help Cognitive Dysfunction?*

The treatment of cognitive dysfunction, which may come and go on its own, is often unsatisfactory. Emotional support and reassurance are important. Corticosteroids are a two-edged sword, and there is little evidence that these drugs alleviate the syndrome when other evidence for active lupus is not also present. I prescribe antimalarial drugs, particularly hydroxychloroquine (Plaquenil), for active lupus and quinacrine (Atabrine) for profound fatigue. Tricyclic antidepressants or serotonin reuptake inhibitors (e.g., Prozac, Zoloft, Paxil) may also be useful, and DHEA appears promising.

## Organic Brain Syndrome

Charlene barely survived an episode of CNS vasculitis when she was 20. At that time, she had high fevers, seizures, and psychotic behavior, and lapsed into a coma for 3 weeks. Now 40 years of age, Charlene is slightly retarded and picks up supplies for Goodwill Industries. Her lupus has not been active in over ten years. Charlene moved to the country to be with her boyfriend. She had not had a seizure in 2 years; therefore, when her epilepsy prescription lapsed, Charlene figured that it was no longer necessary. Last week Charlene had a grand mal seizure at the Goodwill loading dock. Her new doctor looked at her chart and noticed a long-standing history of lupus. He placed her on 60 milligrams of prednisone a day for presumed CNS vasculitis with seizures, but Charlene immediately became agitated and displayed psychotic tendencies. Her old rheumatologist was called on the telephone and explained to her doctor that she had a chronic organic brain syndrome and probably had a seizure due to scarring in brain tissue because of the earlier episode of CNS vasculitis. Charlene was taken off prednisone and resumed her anticonvulsant medication.

In patients who have had a previous stroke because of antiphospholipid antibodies and in those with a history of CNS vasculitis, brain lesions heal with scarring. This results in permanent motor and mental deficits as well as seizures. These patients have what is called *organic brain syndrome*. They do not have active lupus but have scars from previous inflammation related to lupus. Therefore, steroids do not help them and will only increase brain atrophy. Since blood tests and spinal fluid usually reveal that the lupus is inactive, a patient's history is very important. Organic brain syndrome is managed with emotional support and, if needed, psychotropic medications or anticonvulsants.

### Psychosis

Sometimes lupus patients may demonstrate symptoms of psychosis. Psychosis is defined as an inability to judge reality, marked by disordered thinking and bizarre ideas, often including delusions and hallucinations. It usually results in an inability to carry out the ordinary demands of living. The incidence of acute (and fortunately, temporary) psychosis is between 10 and 15 percent during the course of systemic lupus.

Most psychotic episodes occur with CNS vasculitis, but others occur as a result of steroid therapy, water intoxication with low blood levels of sodium, seizures, inadequate antidiuretic hormone secretion, central hyperventilation, or antimalarial therapy. Psychosis may, however, be evident without CNS vasculitis.

Psychosis is managed with corticosteroids when active lupus is evident. I attempt to take a careful history to ascertain the exact use of prescription drugs, over-the-counter medications, street drugs, and herbal or vitamin remedies in order to assess any possible interactions. Antipsychotic preparations (such as Thorazine, Haldol, and Stelazine) are usually used to treat this syndrome. (The management of steroid-related behavior problems is discussed in Chapter 27.)

### Functional Behavioral Syndromes

Depression and anxiety are present in at least half of all lupus patients as a consequence of inflammation, which induces a rapid pulse; cytokines, which may alter mood and behavior; generalized pain, which may result from fibromyalgia; other sources of nonrestorative sleep because of fibrositis or steroids; or inadequate coping mechanisms. These behavioral features of lupus as well as their management are discussed fully in Chapter 25.

## OTHER CENTRAL NERVOUS SYSTEM ABNORMALITIES THAT MAY ACCOMPANY LUPUS

### Complications of Medications Used to Treat Lupus

Agents used to treat lupus can cause CNS symptoms that must be distinguished from what I have described as CNS lupus. The *nonsteroidal anti-inflammatory drugs* are infrequently the cause of headaches and may induce dizziness. Headache and confusion have been reported by 5 to 15 percent of patients taking indomethacin, tolmetin, and sulindac; ibuprofen has, on rare occasions, been associated with aseptic meningitis.

Very high doses of the *antimalarials* (chloroquine, hydroxychloroquine, and quinacrine) have been associated with manic behavior, seizures, and psychosis, while *corticosteroids* can produce agitation, confusion, mood swings, depression, and psychosis. Certain drugs that treat *hypertension* may cause a loss of sexual desire as well as depression. Nearly every prescription medication can affect the CNS.

### Abnormalities Not Related to Lupus Activity

> Michelle had severe active multisystem lupus. Maintained on high doses of prednisone and monthly intravenous Cytoxan, she still had rashes, fevers, swelling, pleurisy, and advanced kidney disease. One day, her fevers rose higher than usual and she began experiencing seizures as well as headaches and a stiff neck. Her doctors hospitalized Michelle for presumptive CNS lupus. An MRI of her brain, however, suggested an abscess. Her spinal tap showed evidence of *Cryptococcus,* which is a fungus not usually seen in healthy people. She was started on antifungal medication, and Cytoxan was discontinued to facilitate the medication's ability to kill the microbes. However, Michelle still required high doses of steroids.

*Infections* of the CNS mimic CNS lupus; they must be carefully considered and ruled out, since lupus patients are especialy susceptible to infection. The most common infectious agents include *Mycobacterium tuberculosis, Meningococcus, Salmonella, Shigella, Staphylococcus,* and *Streptococcus.* Opportunistic organisms are microbes that cause infection only in immunologically compromised individuals, such as those with cancer, patients taking chemotherapy and high doses of steroids, or those who have AIDS. In these people, unusual forms of bacteria, viruses, and fungi can be present. Brain imaging and spinal taps can usually allow doctors to make a definitive diagnosis.

Lupus patients develop *strokes, hypertension, psychiatric disorders, malignancies, aneurysms,* and *Parkinson's disease* at the same or greater frequency as healthy people. However, an established diagnosis of lupus clues the physician into considering other possibilities in evaluating the CNS.

Autoimmune disorders that affect the CNS include *myasthenia gravis* and *multiple sclerosis;* they have an increased incidence among lupus patients. Myasthenia gravis is characterized by rapid muscle fatigue with repetitive tasks, while multiple sclerosis causes blurred vision, loss of bladder and bowel control, as well as difficulty walking. What may add to the confusion is that one-third of multiple sclerosis patients have a positive ANA test. Brain imaging and spinal fluid evaluations usually help differentiate multiple sclerosis from systemic lupus.

Table 10 summarizes the clinical, laboratory, and therapeutic features of some of the important CNS syndromes.

**Table 10.** *Major CNS Syndromes and Their Management*

| Syndrome | Incidence in Lupus (%) | Treatment |
|---|---|---|
| Cerebral vasculitis | 10 | High-dose intravenous steroids; immunosuppressives may be used |
| Antiphospholipid syndrome with brain clots | 5–10 | Platelet inhibitors; anticoagulants |
| Lupus headache | 15 | Migraine therapy, steroids, platelet inhibitors |
| Cognitive dysfunction | 50 | Antimalarials; sometimes steroids; emotional support |
| Chronic organic brain syndrome | 5 | Emotional support, seizure prevention if needed |
| Fibromyalgia | 10–20 | Nonsteroidals, counseling, anti-depressants, physical therapy |
| CNS infection | 1 | Antibiotics |
| Cryoglobulinemia or hyperviscosity | 1 | Steroids, apheresis, chemotherapy, interferon |
| Bleed due to low platelet counts | 2 | Steroids, apheresis, chemotherapy, factor replacement, transfusion |

## NEURODIAGNOSTIC TESTING

A neurologic workup includes a lot more than blood tests and x-rays. Many of the diagnostic techniques used are unique to the CNS. Neurologic testing in lupus patients is divided into several categories, including blood tests, spinal fluid evaluations, brain imaging, electrical studies, and neuropsychological tests.

### Blood Testing

Blood testing is often helpful yet frequently unsatisfactory. It enables physicians to confirm whether lupus is present or active outside the CNS. Inflammation of the brain's blood vessels (called cerebral vasculitis) is usually associated with elevated sedimentation rates, low blood complement levels, and high values for anti–DNA. The presence of antiphospholipid antibodies along with a focal neurologic deficit suggests the antiphospholipid syndrome. Platelet counts should be checked to rule out sources of bleeding. Blood cultures should be obtained if fevers are present. On occasion, it may be important to obtain a serum viscosity or cryoglobulin level. Ribosomal P antibody has a weak association with psychotic behavior and is found only in lupus patients. The finding of antibodies to nerve cells in the blood is also weakly linked with active CNS vasculitis. (These antibodies are discussed in Chapter 11.)

### Spinal Fluid Analysis

A spinal tap or lumbar puncture yields cerebrospinal fluid (CSF). If blood testing does not help make the diagnosis, analysis of spinal fluid becomes important. In CNS vasculitis, spinal fluid often shows a high white cell count, elevated protein levels, LE (lupus) cells, and low sugar values. High levels of antineuronal antibodies may also be found. An immunologic reaction (but not necessarily acute CNS vasculitis) is suggested if oligoclonal bands or increased IgG synthesis rates are found. Patients with the antiphospholipid syndrome usually have normal spinal fluid. If large numbers of red blood cells are present in all the tubes of spinal fluid obtained, bleeding from the brain's blood vessels is suggested. Infections yield positive spinal fluid cultures for bacteria or fungi. Viruses are detected by measuring levels of viral antibodies.

### Brain Imaging

X-rays of the skull are rarely helpful. Isotopic brain scans were used from 1970–1985. Now *computed tomography* (*CT*) *scans* and *magnetic resonance imaging* (*MRI*) are used to reveal strokes, tumors, bleeding, and abscesses. The MRI scans are particularly sensitive and reliable in detecting these conditions. Unfortunately, there has been a tendency to misread any scan abnormality in a patient with known lupus and call it "vasculitis." Focal lesions, or lesions limited to a specific area, suggest the antiphospholipid syndrome, whereas generalized changes are consistent with CNS vasculitis. However, many normal patients and some individuals with inactive disease have minor MRI abnormalities that are difficult to interpret and mean little.

Efforts to study cerebral dynamics, or how the brain works as opposed to the strictly anatomic information derived from CT or MRI, have led to the development of *PET* (*positron emission tomography*) and *SPECT* (*single-positron-emission computed tomography*). PET scanners require a cyclotron and are thus restricted to university medical centers. The major use of such scans is to locate the part of the brain that a seizure is coming from and to show areas of decreased blood flow (hypoperfusion), which suggest cognitive dysfunction.

The "gold standard" for vasculitis remains the *cerebral angiogram,* where dye is injected into the blood vessels of the brain, but this risky procedure is not always positive even in the presence of vasculitis. It may not be necessary if spinal fluid or blood findings are definitive, and the recently developed noninvasive MR angiography can provide nearly as much information.

## Electrical Studies

*Electroencephalograms* (*EEGs*) have been available for decades but are of little help except to identify seizure disorders. Quantitative EEGs and brain mapping studies are more precise for localizing the part of the brain that the seizures are coming from. Multiple sclerosis and lupus symptoms can be localized by using electrical studies with fancy names like *brainstem evoked potentials, visual evoked responses,* and *auditory evoked response measurements.*

*Electromyograms* (*EMGs*) with nerve conduction velocity testing evaluate peripheral nerve problems. They can differentiate a herniated disc from inflammatory nerve or muscle lesions, diabetic numbness, and nerve compressions such as the carpal tunnel syndrome.

## Behavioral Surveys

Several psychological tests have been employed to detect cognitive dysfunction, ranging from the *MMPI* (*Minnesota Multiphasic Personality Inventory*) to the *Luria-Nebraska test,* the *Halsted-Reitan test,* and the *Wechsler Adult Intelligence Scale.* Most neuropsychologists have their own battery of tests customized to the nature of their practice. Although no combination of testing has been validated as being more reliable than any other, these evaluations can help physicians identify depression, psychosis, cognitive dysfunction, and neuroses, among other behavioral disorders.

## Summing Up

The majority of patients with systemic lupus have neurocognitive problems or active inflammation that leads to CNS problems. A wide array of possibilities may account for any given symptom, and a careful workup is necessary in order to avoid inappropriate therapy. The most common complaints include cognitive dysfunction, headache, and fatigue. Manifestations of vasculitis, the antiphospholipid syndrome, and altered behavior are not infrequent. Drugs, infections, and non-lupus-related disorders first have to be ruled out as a cause of the complaint or manifestation. Blood and spinal fluid testing, brain imaging, electrical studies, and neurocognitive evaluations help the physician arrive at a diagnosis. The treatment of CNS lupus can include a combination of anti-inflammatory medications, blood thinners, and emotional support.

# 16
# *The Head, Neck, and Sjögren's Syndrome*

If you are a lupus patient, do you find it hard to see clearly? Do your ears ring? Do you get crops of mouth sores? Do you suck on Lifesavers constantly to make your mouth less dry? Have you ever lost your voice? The eyes, ears, nose, mouth, salivary glands, and larynx are occasionally affected by lupus either as a manifestation of active disease, the antiphospholipid antibody syndrome, or an adverse reaction to lupus medication. Even though they are present in only a minority of cases, pertinent head and neck symptoms and signs are too important to be overlooked. This section reviews the head and neck areas and how problems in these areas are related to lupus.

## HOW DOES LUPUS AFFECT THE OUTER EYES?

There's more to the eye than what we see. Our body's sight organ is divided into several layers, all of which can be inflamed with active SLE. *Discoid lesions* resembling the skin condition called eczema are occasionally observed around the eyelids. The eye muscles may become swollen or inflamed, which is called *orbital myositis,* or temporarily paralyzed due to a *cranial neuropathy,* since 3 of the 12 cranial nerves (peripheral nerves) control the movement of eye muscles. If diabetes and thyroid conditions are ruled out as the cause of orbital pathology, corticosteroids can relieve these conditions.

*Conjunctivitis,* or inflammation of tissues around the eyeball, is more common among lupus patients since they are more susceptible to infection. Antibiotic solutions can keep conjunctivitis in check, but the presence of conjunctivitis does not indicate more active disease. Patients with Sjögren's syndrome (see below) have decreased ability to manufacture tears. When the cornea (the outer covering of the eyeball) cannot be adequately lubricated, it becomes dry and develops pits, leaving scarred areas on its surface. Artificial tears are usually helpful in such cases.

## DOES LUPUS AFFECT SIGHT?

### What about Cataracts, Glaucoma, or Iritis?

The *uveal tract* or middle layer of the eye includes the iris, lens, and ciliary muscles. Lupus does not directly induce *cataracts,* or the clouding of the lens, but corticosteroids can. Similarly, these agents are associated with a greater risk of developing *glaucoma,* which is caused by elevated pressures in the eye. Cataracts can be removed surgically, and eyedrops lower increased ocular pressures.

Inflammation of the iris, or *iritis,* is a recurrent problem in 1 to 2 percent of lupus patients. It has been observed in almost every autoimmune disease and exists as an autoimmune disorder by itself. Iritis is managed with steroid eyedrops, but at times oral corticosteroids, immunosuppressives, or intraocular injections of steroid are needed.

### The Retina

Rachel woke up blind in her right eye. She had a history of mild SLE but had generally felt well. Rachel was married to a stockbroker on the West Coast who was already at work at 6 A.M.; she felt helpless. Michael and she had been trying to have children, and these efforts culminated in two heartbreaking miscarriages. Rachel placed a call to her internist. The internist referred her to an eye doctor. She saw him that day. He performed a fluorescein angiogram that documented a clot in her retinal artery. Rachel was admitted to the hospital, where she was anticoagulated with intravenous heparin followed by oral warfarin (Coumadin). A rheumatologist was called in to see Rachel, and her workup documented evidence for antiphospholipid antibodies. Rachel was advised to take Coumadin for 3 months followed by lifelong low-dose aspirin. Gradually, her sight improved, and within 3 months she had only a small blind spot that did not bother her.

The back layer of the eye is known as the *retina.* It is the only part of the body where physicians can actually see internal blood vessels in their entirety. Although antimalarial drugs or infectious agents infrequently damage the retina, 10 percent of lupus patients develop visible retinal pathology. *Retinal vasculitis,* or inflammation of the retinal vessels, can be painful and cause visual changes. Usually seen with active disease, it is treated with corticosteroids. On the other hand, sometimes areas of "infarction" or dead tissue are noted in the retina. These *cytoid bodies,* as they are called, indicate either old, healed retinal vasculitis or signify that a clot has traveled to the eye as a complication of the antiphospholipid syndrome (Chapter 21). Steroids will not benefit these patients; blood thinning or anticoagulation is the treatment of choice.

When another cranial nerve known as the optic nerve is inflamed, *optic neuritis* results, causing varying degrees of visual impairment. This must be differentiated from blurred vision induced by corticosteroids.

## DOES LUPUS CAUSE HEARING PROBLEMS?

In general, lupus patients have no unusual or specific hearing deficits. They may complain of *tinnitus,* or ringing in the ears, due to anti-inflammatory medications such as antimalarials or NSAIDs. This is especially common in those over the age of 60.

One lupus patient in 500 has a rare complication known as *autoimmune vestibulitis,* characterized by sudden hearing loss with or without visual changes or dizziness. Corticosteroids reverse this deafness.

On rare occasions, the cartilage lining the outer part of the ear becomes inflamed in active SLE, producing a condition known as *chondritis*. It appears identical to another autoimmune disease called *relapsing chondritis,* where cartilage in the trachea, nose, and valves of the heart can also be inflamed. In SLE, however, chondritis is usually limited to the ear and responds to steroids.

To sum up, ear problems are very rare in SLE and, if present, usually stem from nonimmunologic sources.

## THE NOSE

Some of my patients have been wrongly accused of using cocaine when they have nasal septal perforations. The nasal septum—the cartilage membrane that divides the nose into two nostrils—can perforate, or develop holes, in 1 to 2 percent of lupus patients. *Ulcerations* may also be found on the nasal mucosa. I usually prescribe petroleum jelly (e.g., Vaseline) or a steroid preparation (e.g., Kenalog in Orabase) for these. Recurrent *sinus* infections in a patient who does not improve with antibiotics, decongestants, or antiallergy medications and seems to have lupus warrants a blood test for a closely related disorder known as Wegener's granulomatosis. A blood test for anti–neutrophilic cytoplasmic antibody (ANCA) can help differentiate these two types of vasculitis, which are treated quite differently.

## DENTAL CONSIDERATIONS

Very few patients realize the importance of good oral hygiene as an adjunct in managing SLE. Since the mouth is teeming with bacteria, some rheumatologists believe that all dental procedures should be accompanied by prophylactic antibiotics to protect patients with lupus from developing infected heart valve vegetations (Chapter 14) or other types of infections. Those who have temporomandibular joint (TMJ) involvement from SLE or oral scleroderma-like tightening of the lips with decreased opening of the jaws find chewing more difficult and may develop cavities and abscesses. They should see their dentist at least once a year for a checkup.

Sores in the mouth, or oral ulcerations, are seen in 20 percent of lupus patients and are discussed in Chapter 12. Sjögren's syndrome causes dryness of the mouth and is discussed below.

## WHAT IS THE REASON FOR THE RASPY VOICE?

The voice box (vocal cords) is in the larynx. A synovially lined joint known as the cricoarytenoid joint is found in this area (Chapter 13). Lupus activity sometimes produces synovitis. When the synovium of the cricoarytenoid joint becomes inflamed as a result of active disease, a hoarse voice can develop. Ear, nose, and throat specialists see evidence of this when they perform direct laryngoscopy. Anti-inflammatory medications can relieve the problem, although sometimes the joint may be sprayed with a steroid aerosol or locally injected.

## SJÖGREN'S SYNDROME

### Do The Skin and Eyes Feel Dry?

Laura was unable to wear the contact lenses her optometrist had prescribed. A fashion model, Laura tried her best not to think about her lupus, but fortunately the condition was limited to her skin (which could be covered with hypoallergic makeup) and symptoms of occasional aching and fatigue. The only other medicine she took was Naprosyn for intermittent joint pain. The optometrist referred her to an ophthalmologist, who diagnosed Sjögren's syndrome. She prescribed artificial tears and told Laura to drink plenty of fluids. When Laura told this to her rheumatologist, he added Plaquenil and gave her literature from the Sjögren's Foundation, which listed over a hundred over-the-counter moisturizing agents for the ears, nose, mouth, eyes, and vagina. Laura's joint aches and fatigue improved, but she still had to use moisturizing agents.

### What Is Sjögren's Syndrome?

In 1933, Henrik Sjögren (pronounced show-gren) described some symptoms common to a group of patients: dry eyes (keratoconjunctivitis sicca), dry mouth (xerostomia), and arthritis. In the late 1960s, doctors found that many patients with these symptoms had an autoimmune process. Sjögren's syndrome, as the disorder is now known, can be part of many autoimmune diseases or may exist by itself (termed *primary Sjögren's*) without fulfilling accepted criteria for any other process. It is estimated that 5 to 10 percent of all Sjögren's patients have SLE, although the incidence of Sjögren's in established SLE patients has not been reliably ascertained. We do know that at least 10 percent of the individuals with lupus have obvious Sjögren's. However, if minimally or asymptomatic patients with lupus undergo vigorous testing for the syndrome, perhaps as many as one-third would fulfill accepted Sjögren's definitions.

## How Do We Diagnose Sjögren's and Why Is It Important?

Some of the procedures used to diagnose Sjögren's include the *Schirmer's test,* which measures the amount of tearing, or *Rose Bengal staining* of the cornea, which looks for pitting or areas of scarring. Although simple to perform, these procedures have false-positive and false-negative results. The ultimate and most accurate diagnostic test is a *lip biopsy,* which displays a characteristic inflammatory infiltrate. Fortunately, it is rarely necessary to do this potentially painful procedure.

Why is it important to diagnose Sjögren's? First of all, the syndrome is associated with disease outside the salivary glands and tear ducts. For example, dry lungs or *bronchitis sicca* predisposes one to interstitial lung disease (Chapter 14). Other related disorders include atrophic gastritis (dry stomach), hyperviscosity syndrome (thick blood), cryoglobulinemia, and subacute cutaneous lupus rashes. The anti-Ro (SSA) antibody (Chapter 11) is noted in 30 percent of those with SLE but 70 percent of those with primary Sjögren's. This antibody crosses the placenta and has the capability of inducing neonatal lupus and congenital heart block (Chapter 22). Of all rheumatic diseases, Sjögren's correlates with the highest levels of autoantibodies. It is not unusual for Sjögren's patients to have ANA tests or rheumatoid factors with levels well over several thousand. Sjögren's is associated with the HLA-DR3 marker (Chapter 7). Also, Sjögren's syndrome is the only autoimmune condition that has a potential for malignant transformation. Up to 5 percent of Sjögren's patients develop a blood disorder, particularly lymphoma.

## How Should Sjögren's Be Managed?

In spite of all that has been said, Sjögren's syndrome is usually a relatively benign process and its treatment is symptomatic. For instance, dry eyes usually respond to artificial tears and dry mouth to everything from Lifesavers to sparkling water with a touch of lemon, lime, orange, grapefruit, or any other citrus fruit (although frequently used, Lifesavers are discouraged because of their high sugar content). A humidified environment or room humidifiers can be helpful. Chicken soup moisturizes the lungs, as does a drug called Humabid. Recently, some evidence has suggested that hydroxychloroquine (Plaquenil) helps to mitigate the underlying immune process of Sjögren's. Pilocarpine increases tearing and promotes increased salivary secretions. Most experiences with prescribing corticosteroids for Sjögren's have been disappointing except when they are used for a short time in a rare subset of Sjögren's called *Mikulicz's syndrome* where the parotid or salivary glands become greatly enlarged from acute inflammation. This condition is readily identified, since patients appear to have the mumps, and the salivary glands are extremely tender to the touch.

## Summing Up

The most common cause of head and neck involvement in SLE is Sjögren's syndrome. Eye complaints are often due to medication, especially steroids, but 10 to 15 percent of patients with lupus develop complications from disease activity or the antiphospholipid antibody. Ulcers of the mouth or nose are seen in 20 percent of these patients. Involvement of the ear or larynx is rare. Patients and their physicians must carefully evaluate head and neck complaints, since failure to intervene promptly with autoimmune inflammation of the optic nerve or the ear can result in permanent blindness or deafness.

# 17
# *What about Hormones?*

Why are women disproportionally afflicted with SLE? Do men have milder or more severe cases than women? Are there hormonal imbalances unique to lupus? Is there any relationship between the disease and our endocrine (hormonal) tissues, such as the thyroid and adrenal glands? Since it was first observed that 90 percent of lupus patients are women and 90 percent of these patients develop the disease during their childbearing years, it seemed logical to study hormonal relationships in SLE. This section reviews these approaches.

The use of hormonal interventions such as birth control pills and estrogens, among others, are discussed in Chapters 13 and 28.

## UNDERSTANDING HORMONES

A hormone is a chemical made by one organ that is transported in the blood to another organ, where it carries out its function. There is an area in the brain known as the hypothalamus, which is responsible for manufacturing a group of chemicals known as *releasing hormones.* When these chemicals travel a short distance to another area called the anterior pituitary gland, the releasing hormones promote the production of *stimulating hormones.* These stimulating hormones are secreted into the bloodstream and travel to the peripheral endocrine glands or other body organs.

For example, the hypothalamus makes thyroid-releasing hormone, which induces pituitary production of thyroid-stimulating hormone, which, in turn, prompts the thyroid gland to make thyroid hormone. Other hormones derived by similar mechanisms include cortisol, made by the adrenal gland, and prolactin, which is made in the anterior pituitary and results in the secretion of breast milk. The female reproductive organs produce estrogen and progesterone; male hormones are called *androgens,* with testosterone being the most important. Figure 9 illustrates these hormonal pathways.

## HOW IS THE ENDOCRINE SYSTEM ALTERED IN SYSTEMIC LUPUS?

What happens to sex hormones in lupus patients? Sex hormones act on the immune system in three ways. First of all, they stimulate the central nervous

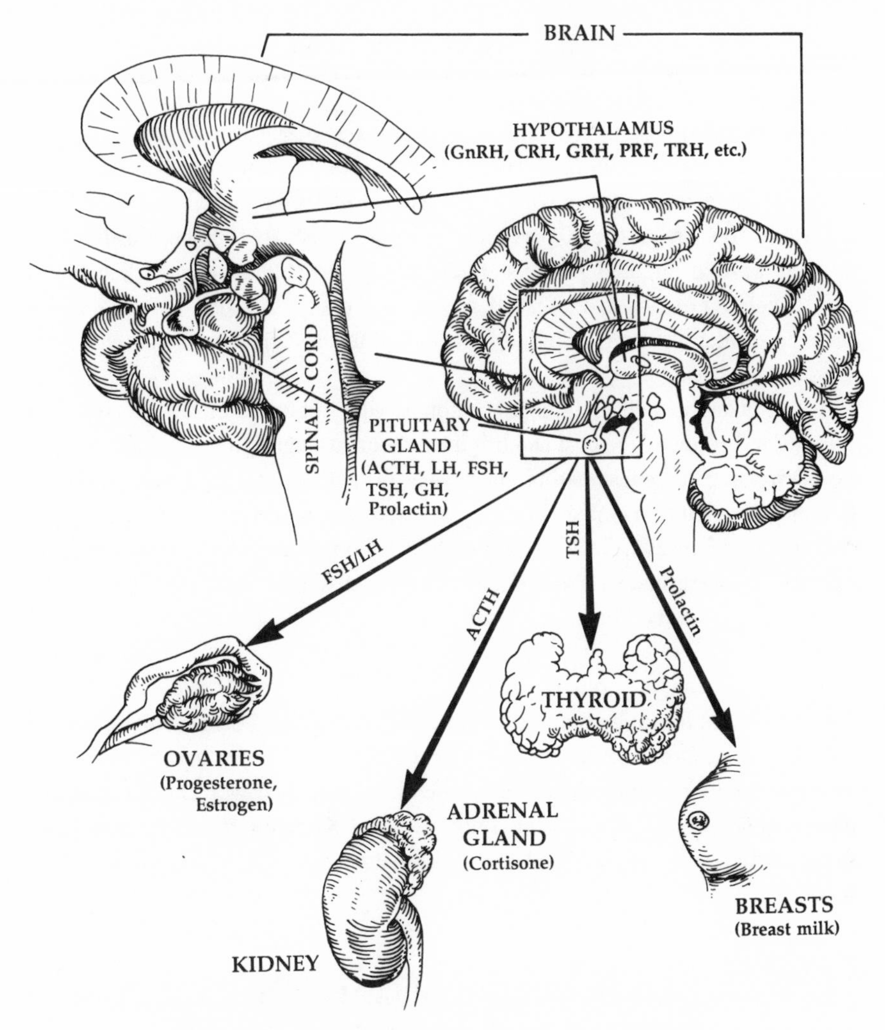


**Fig. 9.** *Hypothalamic and Pituitary Hormonal Pathways*

system to release immunoregulatory chemicals. Second, they regulate the production of cytokines (Chapter 5). Finally, sex hormones stimulate endocrine glands to release other hormones, such as prolactin, in women.

Estrogens, or female hormones, can promote autoimmunity, and this can indirectly increase inflammation, whereas androgens (male hormones) generally suppress autoimmunity. Estrogens increase the production of autoantibodies, inhibit natural killer cell function, and induce atrophy of the thymus gland. Further, in SLE, estrogens are metabolized differently. Due to an abnormality in a chemical pathway (called 16 alpha-hydroxylation), lupus patients have excess levels of 16 alpha-hydroxyestrone and estriol metabolites. Males with lupus have lower-than-normal levels of testosterone and other androgens.

## IS THERE A DIFFERENCE BETWEEN MEN AND WOMEN WITH SYSTEMIC LUPUS?

How do males with SLE fare? This information has been surprisingly difficult to ascertain. For instance, in order to study outcomes in a scientifically acceptable fashion, a large group of men with SLE must be followed closely for at least 5 and preferably 10 years. Since 90 percent of those with SLE are women, this requires studying several hundred patients. Interestingly, no studies suggest that males have a better prognosis. Published reports are evenly divided between males having a similar outcome to that of females and those suggesting that men have a poorer prognosis.

If SLE is aggravated by female hormones, why might males fare worse? This question has stumped the best and brightest rheumatologists for years. One possible answer recalls a fascinating survey performed in the 1960s that would be extremely difficult to repeat in today's research environment. In that report, the gender of fetuses from women with SLE who miscarried was tabulated, and the majority were males. This suggests that male fetuses with a lupus gene are more likely not to be born, which may explain why there are so few men with SLE.

On the other hand, a rare chromosomal disorder known as *Klinefelter's syndrome* provides investigators with contradictory clues. A normal female carries two X chromosomes (XX) and a normal male one X and one Y (or male) chromosome (XY). Individuals with Klinefelter's are men who carry an extra X chromosome (XXY). It has been suggested that Klinefelter's patients have an increased incidence of SLE and that this is directly related to female hormone excess.

## MENSTRUAL PROBLEMS IN SLE

One of the most frequently asked questions I encounter deals with the relationship between lupus and menstruation. *Amenorrhea,* or the absence of menstruation, is observed in 15 to 25 percent of women with SLE between the ages of 15 and 45 and can be related to prior chemotherapy or severe disease activity. Irregular periods are not uncommon. For example, nonsteroidal anti-inflammatory drugs (NSAIDs) increase bleeding, and changes in corticosteroid doses or steroid shots into joints or muscles can alter menstrual cycles. A pregnancy test should always be performed before any medical treatment directed toward amenorrhea or menstrual irregularities is initiated.

*Premenstrual syndrome* (*PMS*) may be more severe in SLE, and the majority of women with lupus report a mild premenstrual flare in their musculoskeletal symptoms. The onset of *menopause* is associated with less SLE activity.

## THYROID DISEASE

Vanessa was feeling achy and under a lot of stress. When she visited her doctor, he decided to take an ANA test, which came back positive, and he told her she might have lupus. She was very edgy and couldn't focus on any task for more than 3 minutes. She increasingly became aware of palpitations and was always turning up the air conditioning. A rheumatologist was consulted, who confirmed that the ANA was positive, but it was in a low-level speckled pattern. The blood panel also included other autoantibodies, and her thyroid antibodies were markedly elevated. Blood testing showed that she had elevated thryoid function and was hyperthyroid. A diagnosis of Hashimoto's thyroiditis was made; Vanessa did not have lupus. When antithyroid medication failed to suppress her symptoms adequately, she was given radioactive iodine to drink, which prevented the gland from making thyroid and allowed her to feel normal again. She is now maintained on a low dose of thyroid replacement.

The thyroid gland in the neck helps regulate our metabolism and it affects how we feel by controlling the production of thyroid hormone. Thyroid-related symptoms—such as fatigue, palpitations, fevers, being too hot or too cold, and joint aches—can often be mistaken for SLE.

Autoimmune disease of the thyroid is characterized by detectable levels of antithyroid antibodies in the blood. Clinically manifested as *Graves' disease* or *Hashimoto's thyroiditis,* autoimmune thyroid disease initially appears as hyperthyroidism (overactive thyroid) and ultimately develops into hypothyroidism (underactive thyroid). Approximately 10 percent of lupus patients have thyroid antibodies, and autoimmune thyroiditis occasionally coexists with SLE. Many of these individuals also have Sjögren's syndrome. Conversely, many people like Vanessa with primary autoimmune thyroid disease have positive ANA tests without evidence of lupus.

Whether or not autoimmune thyroiditis is present, some 1 to 15 percent of those with SLE studied at any point are hyperthyroid and 1 to 10 percent are hypothyroid. In other words, thyroid abnormalities are commonly noted in lupus and are related to antithyroid antibodies about half of the time.

## DIABETES MELLITUS

The most common cause of diabetes in SLE patients is corticosteroid therapy. Steroids raise blood sugars, and the risk of developing diabetes depends on how much prednisone the patient has been taking and for how long. Diabetes sometimes disappears when steroid doses are lowered or discontinued.

The islet cells of the pancreas are endocrine tissues responsible for producing insulin. Interestingly, many individuals with juvenile diabetes (called type I) have positive ANAs even though lupus is uncommon in this group of patients.

## THE ADRENAL GLAND

*Corticotropin-releasing hormone* (*CRH*) in the hypothalamus stimulates the production of *adrenocorticotropic hormone* (*ACTH*) in the pituitary, which induces the secretion of various types of cortisone (steroids) by the adrenal gland (see Figure 9). The adrenal gland can be affected directly and indirectly in systemic lupus.

The direct effects include an autoimmune inflammatory process known as *autoimmune adrenalitis,* whose concurrence rate with SLE has not been determined. In addition, the *antiphospholipid syndrome* (see Chaper 21) may result in clots to the adrenal gland's blood supply, which leads to infarction, or death of adrenal tissue.

The administration of oral (exogenous) corticosteroids suppresses internal (endogenous) secretions of steroids from the adrenal gland. The normal adrenal gland makes the equivalent of 7.5 milligrams of prednisone daily. Higher oral steroid doses turn off the adrenal gland; when steroid doses are decreased below 7.5 milligrams, the gland is supposed to resume steroid production. However, when one has been on corticosteroids for a prolonged time, the adrenal response is sluggish and symptoms of adrenal insufficiency, such as fatigue and aching, may become evident. When the adrenal gland does not make enough cortisone, symptoms of *adrenal insufficiency* become manifest and can mimic a lupus flareup. See Chapter 27.

## WHAT ABOUT PROLACTIN—THE BREAST HORMONE?

*Prolactin* is a hormone that provides the stimulus needed to produce breast milk. Over the last few years, several investigators have documented that a disproportionate number of lupus patients have elevated prolactin blood levels. One group of researchers has gone further and suggested that a drug known as bromocriptine (Parlodel), which blocks prolactin, may help lupus patients. These hypotheses have not yet been confirmed, and lupus patients of mine who have taken the drug have not improved.

## SEXUAL DYSFUNCTION: DO LUPUS PATIENTS HAVE A PROBLEM?

It's rare for women with lupus to be unable to enjoy sexual intercourse. Female sexual dysfunction because of disease activity is unusual in SLE. The two well-established exceptions are vaginal dryness from Sjögren's syndrome and vaginal ulcerations, which are rare compared to oral or nasal ulcerations. Both of these conditions can result in painful intercourse. Vaginal dryness is managed with lubricants and ulcerations with a hydrocortisone ointment. Physiologic sexual problems unique to males with lupus have not been reported.

Sometimes women with avascular necrosis or other destructive changes in the hip have difficulty spreading their legs apart during lovemaking. Until corrective surgery can alleviate this, lupus patients and their partners are urged to use alternatives to the missionary (man on top) position, such as the woman being on top, or rear entry. (In Chapter 25, psychological and emotional aspects of sexuality are discussed in more detail.)

## DO LUPUS PATIENTS HAVE ANTIBODIES TO SEX HORMONES?

Even though sexual problems of a physiologic nature are unusual, many lupus patients have antibodies to reproductive hormones. As many as 30 percent of women with lupus have antiestrogen and antiovarian antibodies in their blood. The presence of the antibodies does not really matter, and has nothing to do with fertility, except that they are more common in patients taking birth-control pills. Males with SLE have an increased prevalence of antisperm antibodies. Again, this has no known clinical relevance.

## Summing Up

Hormones are substances secreted by an endocrine gland in response to stimulation by the brain. They are capable of modulating immunologic reactions, but the extent to which this is clinically useful is not known except that female hormones tend to promote immune responses, whereas male hormones are more immunosuppressive. Nevertheless, male lupus is often more severe than female lupus. Patients with SLE have an increased incidence of antibodies to glandular tissue and autoantibodies to reproductive organs, but the significance of this is not known. Sexual dysfunction is uncommon except that a few women with SLE have disease-related vaginal dryness, vaginal sores, or arthritic hips.

# 18

# *The Impact of Lupus upon the GI Tract and Liver*

The largest organ system in our body is the gastrointestinal (GI) tract. If it were laid out on a floor, the area from the throat to the anus would extend for 40 feet. So it should come as no surprise that such an important part of the body can be involved in lupus. Actually, it is surprising that only limited portions of the gastrointestinal system are involved in the disease. This chapter will cover the relevant information for the lupus patient.

In addition, the GI system includes closely related regions that are not part of the GI tract *per se* but either empty into it or have closely interrelated functions. These areas consist of the liver, pancreas, and biliary tree (bile ducts and gall bladder). In order to lend some order to this section, we will start at the top and work down. Figure 10 highlights the anatomy of the GI tract.

## TAKE A GULP!

The upper GI tract begins in the throat and ends just beyond the stomach. Starting from the top, persistent sore throats are common complaints of children with SLE, even though nothing is usually found on physical examination. (Mouth sores are discussed in Chapter 12 and dental hygiene in Chapter 16.)

The oral cavity gives way to a long tube called the esophagus, which carries food and liquid nourishment from the mouth, behind the chest cavity and heart, and finally to the stomach. Esophageal problems in SLE are of two types: those related to muscle dysfunction and symptoms related to reflux, or heartburn. The upper part of the esophagus wall contains a type of muscle called *striated* and the lower esophagus a type of muscle called *smooth*. The upper muscles are responsible for swallowing. Patients with certain other rheumatic diseases such as inflammatory myositis (dermatomyositis or polymyositis) have a high incidence of *dysphagia,* or difficulty swallowing. Interestingly, they seem to tolerate solid foods quite well, but liquids may come up through the nose and are sometimes aspirated into the lungs. Approximately 10 percent of patients with lupus have a crossover (mixed connective tissue) disease with components of inflammatory myositis.

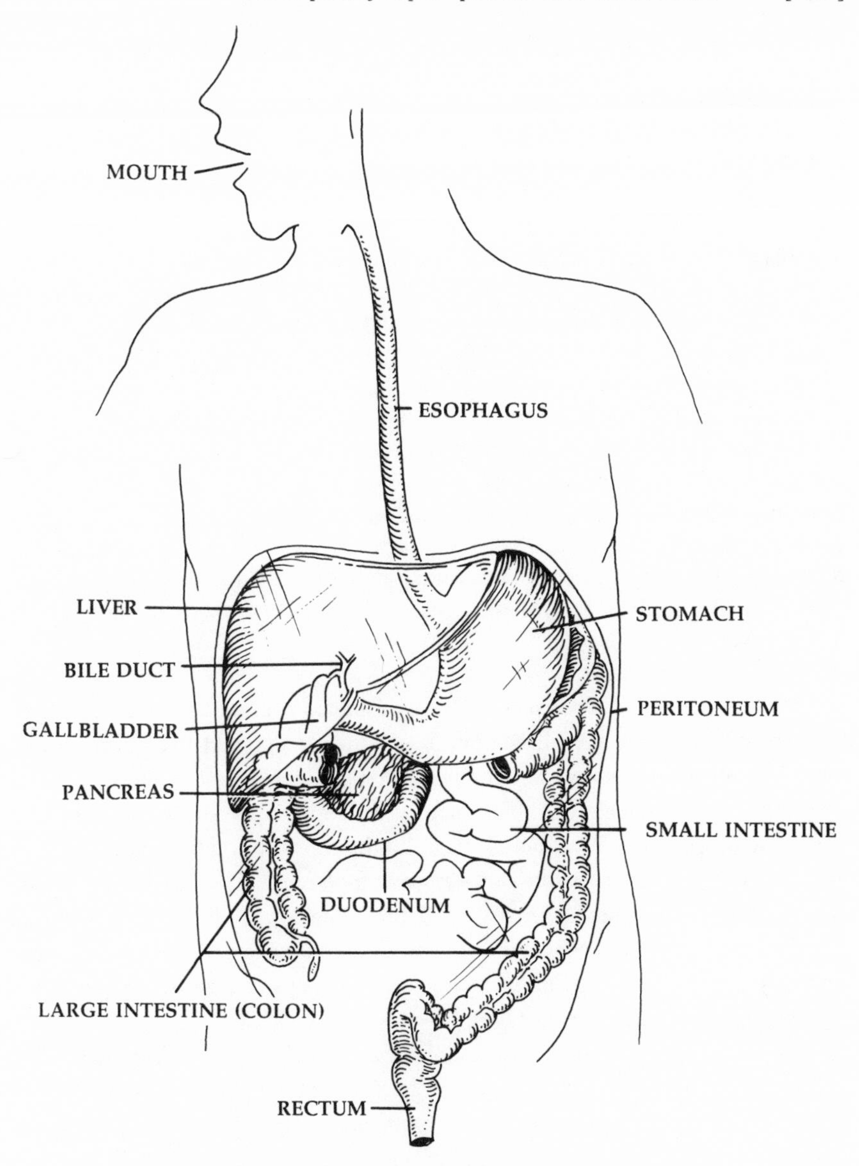


**Fig. 10** *Functional Anatomy of the GI Tract*

Barium swallows (a limited upper-GI x-ray procedure) or esophageal manometry (which measures pressures) easily identifies the syndrome. About once a year, I encounter an individual whose swallowing problem has posed such a significant threat of aspiration pneumonia or malnutrition that I have one of my colleagues in head and neck surgery perform a procedure on the muscles known as a cricopharyngectomy, which usually eliminates the problem.

## WHY DO LUPUS PATIENTS GET HEARTBURN?

Phillip always had a sensitive stomach. When he started treatment for lupus, his doctor prescribed Naprosyn, which relieved his joint pain but gave him a mild gastritis. Phillip had a long-standing history of heartburn, which worsened around the time he was diagnosed. An upper-GI x-ray series was normal except for evidence of a hiatal hernia (fluid going back into the esophagus from the stomach from displacement). He tolerated the gastritis and didn't tell anybody about it. Three months later, he attended a Bruin football game. After an afternoon of fun and frolic which included eating the end-zone pizza, consuming two Bruin dogs, and drinking three beers, Phil began to feel as if he was digesting this haute cuisine again and again. He thought that if he took an extra Naprosyn it might help his stomach pain. The next morning he passed black, tarry stools and nearly passed out. His family took him to an emergency room. When blood was drawn, he was markedly anemic with a hemoglobin of 6. Dr. Gordon took him to have an endoscopy (where doctors can visualize the stomach), where Phil was diagnosed as having a hiatal hernia, erosive gastritis, and a duodenal ulcer. Lucky to be alive, Phil was given a transfusion, started on Prilosec, taken off Naprosyn, and given an appointment with the nutritionist.

Ten to 50 percent of lupus patients have problems with the esophagus, making them painfully aware that this organ is not lined to protect itself from "acid rain." Esophageal tissue is not equipped to handle acid, but stomach tissue is specially lined to handle this. The esophagus empties into the stomach, which makes acid. There are a couple of reasons why fluid from the stomach ends up back in the esophagus, or produces *reflux esophagitis.* It could be the way a person is built. The stomach may be higher than usual, and the sphincter (a muscle) between the esophagus and stomach is weakened. This anatomic anomaly is a common finding in both healthy patients and particularly patients with lupus; it is called a *hiatal hernia.*

It could also be that your lower esophageal muscles may be incompetent and fail to propel materials into your stomach correctly. Patients with crossover or mixed connective tissue disease have components of scleroderma and/or myositis and are especially predisposed to these problems. The net result of acidic stomach fluid entering the esophagus is not only heartburn but also scarring, adhesions, and even poorer peristalsis (propulsion of the lower esophagus) from chronic "burns."

An upper-GI x-ray series is usually diagnostic. On occasion, an upper-GI endoscopy identifies additional problems, and at this procedure the esophagus can be dilated, which may help to relieve symptoms. With heartburn, people often automatically reach for antacids, which is a natural and understandable reflex. But it's only part of the solution. Esopahageal motility problems are managed by eating small, frequent meals rather than a few large ones, and patients are generally instructed not to lie down for at least 2 hours after eating. Antacids, $H_2$ antagonists (Tagamet, Zantac, Axid, Pepcid), motility agents (Reglan, cisapride or Propulsid), sodium sulcrafate (Carafate), and especially omeprazole (Prilosec) are all effective to varying degrees.

## INTERNALS AND EXTERNALS: NAUSEA, VOMITING, DIARRHEA, AND CONSTIPATION

Anita experienced a change in her bowel habits. After a particularly stressful experience at work, she began having diarrhea alternating with constipation with mucousy stools. This was associated with bloating, cramping, and distension. After a complete GI workup that ruled out other possibilities, Dr. Berk diagnosed her as having functional bowel disease. Around the same time, Anita's internist told her that the muscle and joint aches she complained about in her upper back and neck areas and which were keeping her up at night were due to a closely related disorder known as fibromyalgia. Well-controlled on Levsin and a high-fiber diet, even Anita's fibromyalgia improved. Then she was diagnosed with SLE. Wondering whether lupus had affected her intestines all along, Anita requested and her doctor performed another evaluation which showed no evidence of lupus activity in the GI tract. Dr. Berk diagnosed her as having a flareup of functional bowel syndrome, which was resulting from generalized lupus activity.

At various times, many of my patients complain of nausea, vomiting, diarrhea, or constipation. The most common sources of these complaints are nonsteroidal anti-inflammatory drugs (NSAIDs), corticosteroids, and chemotherapy. The rest of the bowel complaints listed here are what doctors label *functional.* In other words, they represent a disorder that carries several names including *functional bowel disease, spastic colitis,* or *irritable bowel syndrome*—all of which are related to abnormal intestinal muscle function. This condition correlates with the presence of fibromyalgia, which is not part of SLE. It is, however, seen in a large number of lupus patients. Occasionally, the symptoms of nausea, vomiting, diarrhea, or constipation reported by SLE patients are due to lupus or a concurrent inflammatory bowel disease such as ulcerative colitis.

## PEPTIC ULCER DISEASE

Have you ever experienced a pain in the pit of your stomach that would not go away? Until 1975, 20 percent of patients with SLE developed an ulcer during the course of their disease. With the introduction of medications such as the $H_2$ blockers (Tagamet, Zantac, Pepcid, Axid), sodium sucralfate (Carafate), omeprazole (Prilosec), as well as an agent known as misoprostol (Cytotec), the number of lupus patients with ulcers has decreased to less than 5 percent. Though necessary to treat the disease, NSAIDs and corticosteroids can all produce erosions in the stomach or duodenum of the small intestine, leading to ulcerations. Fortunately, recently introduced NSAIDs (Lodine, Relafen) and nonacetylated aspirin products (Disalcid, Trilisate) are much less likely to cause an ulcer than the older nonsteroidals.

## IS INFLAMMATORY BOWEL DISEASE ASSOCIATED WITH SLE?

Somewhere between 1 and 4 percent of lupus patients experience severe, crampy abdominal pain with chronic diarrhea. They have developed a second autoimmune process known as *ulcerative colitis*. Characterized by inflammation of the superficial lining of the colon (large intestine), ulcerative colitis is treated with aspirin/sulfa antibiotic combinations (e.g., Azulfidine) and, if necessary, steroids. Patients with lupus may have difficulty tolerating sulfa derivatives.

*Crohn's disease* (or *regional ileitis*) and SLE are both autoimmune diseases, but curiously, very few patients are victims of both disorders. Since both are autoimmune diseases, an increased concurrence with SLE would be anticipated. In fact, lupus and Crohn's disease have been reported together only a few times in the world's literature.

## WHAT IS ASCITES?

Some lupus patients notice a swelling in their belly and feel as though they were pregnant. It has been estimated that at some point 10 percent of lupus patients demonstrate *ascites,* a collection of fluid made by peritoneal tissue. The peritoneum is a thin membrane that lines the abdominal cavity, just as the pleura and pericardium line the lung and heart. We have already shown how pleurisy and pericarditis develop as a result of irriation of these linings (Chapter 14). Peritonitis evolves similarly. Irritation of the peritoneum results in fluid formation (ascites) throughout the abdominal area. Ascitic fluid, like pleural fluid, is either a transudate or an exudate.

*Transudates* (clear, sterile fluids) are painless and common when pleural or pericardial effusions are also present. Nephrotic syndrome, a condition observed when the kidney leaks protein, is also associated with ascites. *Exudates* can be painful and are thicker and cloudier, producing ascites when there is an infection, malignancy, pancreatitis, or serious inflammatory process in the abdomen. Usually identified at physical examination, ascites is also diagnosed by an ultrasound or computed tomography (CT) scan.

When ascites is diagnosed, I usually arrange to have some of the fluid removed for microscopic analysis and culture to help identify its source and cause, which, in turn, suggests appropriate management. Exudative ascites can be mistaken for a "surgical abdomen," resulting in unnecessary surgery. If an infection is ruled out, ascites is treated with anti-inflammatory medication, gentle water pills (diuretics), and occasionally periodic drainage. Infections are treated with antibiotics.

## MALABSORPTION

Have you ever felt that the food you eat isn't getting into your system? Severe diarrhea with very low serum protein (especially albumin) levels is rare in SLE, but when it does occur, it should tip off the lupus specialist that protein is being lost through the intestine as a consequence of malabsorption. Abdominal pain may also be present. Among those who suffer from malabsorption, 90 percent are children; less than 1 percent of the adults with SLE malabsorb their food. *Protein-losing enteropathy,* as this symptom complex is called, is treated with corticosteroids.

## LESS FREQUENT COMPLICATIONS: MESENTERIC VASCULITIS, INFARCTION, AND BOWEL HEMORRHAGE

> Jackie had a severe case of lupus, being treated with 40 milligrams of prednisone a day, which her doctor wanted to raise. Jackie, however, had become diabetic and developed high blood pressure already. Since she was still getting used to having lupus, she was not ready for the increase in medication. One morning, she developed a fever, severe abdominal pain, and bloody diarrhea. Her doctor hospitalized Jackie and called in a surgical consultant. The diagnosis of an acute abdomen was made and she was taken to surgery. In surgery, her doctors discovered vasculitis along with a small perforation that resulted from steroid use. Nutrition became a major problem, and peritonitis set in that seemed resistant to all antibiotics. Her situation is critical and her doctors are not sure she will make it. They are hopeful that an experimental approach will help.

If cramping and abdominal pain are associated with vomiting, fever, and bloody stools, an immediate call to the doctor is in order. *Mesenteric vasculitis,* or inflammation of the blood supply to the small and large intestines, is one of the most serious complications of lupus. It calls for urgent intervention, since it is life-threatening. Even in the best of hands, this condition carries up to an 80 percent mortality rate. Estimates suggest that 1 to 3 percent of lupus patients may develop this complication. Patients on steroids are especially at risk, since the hormone thins the lining of the bowel and makes it more susceptible to perforation.

The blood supply to the bowel can also sustain blood clots in patients with antiphospholipid antibodies (Chapter 21), which also leads to *mesenteric infarction*. This implies that not enough oxygen is getting to the bowel tissue because of an interruption in the blood supply. Both mesenteric vasculitis and antiphospholipid antibodies can induce mesenteric infarction.

Since the body cannot survive for more than a week or two with infarcted or "dead" bowel, surgical removal of this tissue is necessary. The best way for a doctor to approach mesenteric vasculitis or infarction is to be on the lookout for it. Few rheumatologists who regularly treat lupus patients see this complication more than once every 2 to 3 years. Additionally, if vasculitis is evident, costicosteroids are given; if a clot is present, the patient is given anticoagulants after surgery.

## THE BILIARY TREE

Bile is formed in the liver, stored in the gallbladder, and drained into the duodenum of the small intestine through bile ducts. Systemic lupus rarely affects this region except for the pancreas. The pancreas is a secretory and endocrine organ. It makes digestive enzymes that help the intestine break down food and empty it into the duodenum through the biliary tree. As an endocrine organ, the pancreatic islets manufacture insulin.

## WHAT ABOUT THE PANCREAS?

Wendy thought she would die. Although her lupus was under reasonably good control with steroids and Imuran, she never knew back pain could be as bad as this was. It occurred suddenly after an evening of drinking beer with her college roommate, who had just broken up with Wendy's former boyfriend. At 3 A.M., Wendy noticed the sudden onset of searing back pain and knew she had overdone it. Her roommate took her, screaming in pain, to the emergency room, where her serum amylase was 2000 and a diagnosis of acute pancreatitis was made. Dr. Dietz was not sure if the pancreatitis was from her prednisone, Imuran, alcohol abuse, or lupus. Her rheumatologist was asked to see Wendy. She stopped all medication except for intravenous steroids; she also ordered intravenous fluids, pain medicine, and nasogastric suction. It took a week for Wendy to start coming around, and the culprit was not active lupus but either Imuran (which was not resumed) or life-style habits. Wendy was told that she could not have any alcohol at any time for the next few years, and she has adhered strictly to this regimen.

I hope you never experience the severe midabdominal pain radiating to your back that Wendy had, which is associated with nausea, vomiting, and fever. In lupus, several mechanisms contribute to this extremely painful form of pancreatic inflammation known as *pancreatitis*.

Certain agents used in the management of lupus have been found to provoke pancreatitis. These include corticosteroids, azathioprine (Imuran), and thiazide diuretics (e.g., hydrochlorothiazide, Dyazide). Vasculitis of the pancreatic blood supply is the second most common cause of pancreatitis in SLE. One can also develop pancreatitis for the same reasons that otherwise healthy people do—alcohol abuse, gallstones, and physical trauma to the back where the pancreas is.

Ultrasounds or CT scans are obtained to look for gallstones or "cysts" in the pancreas (pancreatic pseudocysts), which could be perpetuating the problem.

The management of pancreatitis is quite problematic. Since pancreatic vasculitis is treated with steroids and steroid-induced pancreatitis is treated with steroid withdrawal, this serious process is first approached with general measures until the cause is determined. These include giving you nothing by mouth, supplying intravenous hydration, putting a tube through your nose that goes into the intestine to remove secretions, and administering pain medicine. If previously prescribed, thiazide diuretics and azathioprine are discontinued. If generally active lupus is evident, steroid doses can be briefly but greatly increased to treat presumed pancreatic vasculitis. Should steroids be found culpable, they are tapered but cannot be abruptly discontinued. In spite of aggressive measures, recurrence is common and the process can go on for months.

## HOW DOES LUPUS AFFECT THE LIVER?

How can you tell if your liver is involved? Most of the time, no symptoms or signs are evident until advanced disease is present. On occasion, right-sided upper abdominal pain of distension, fevers, or a yellow appearance are clues. Lupus activity in one form or another in the liver is evident in numerous ways that are delineated below.

Involvement of the liver in SLE is a frequently misunderstood complication of the disease. It can be affected as a result of both lupus and medications used to treat inflammation. The concepts of what constitutes *autoimmune (lupoid) hepatitis* has undergone many changes since it was first described in the 1950s. This section attempts to reconcile our differing perceptions of what "lupus in the liver" really means.

*Enlargement of the liver,* or hepatomegaly, is found in 10 percent of patients with SLE. The liver is rarely tender unless the enlargement is so great that the capsule or covering of the organ is stretched. The most common causes of large livers in lupus include autoimmune hepatitis, ascites, congestive heart failure, or a complication of a large spleen, whose materials drain into the liver.

*Jaundice,* the condition we think of as turning a patient yellow, is seen in 1 to 4 percent of patients with lupus. Manifested by high serum levels of bilirubin, which are responsible for that yellow pigmentation and itching, jaundice results from autoimmune hemolytic anemia, viral hepatitis, cirrhosis, or bile duct obstruction from gallstones, tumor, or pancreatitis. Occasionally, certain medications, including NSAIDs and azathioprine, may produce jaundice.

*Hepatic vasculitis,* or inflammation of the small and medium-sized arteries of the liver, is extremely rare and is noted in just one lupus patient per thousand. It responds to corticosteroids.

*Budd-Chiari syndrome* results from a blood clot in the portal veins, which drain materials from the liver into the vena cava. Those patients with antiphospholipid antibodies appear to be uniquely at risk for developing these clots. Additionally, hepatic artery clots may occur. Untreated Budd-Chiari syndrome can lead to ascites, elevated liver pressure (called portal hypertension), and liver failure. The preferred treatment of Budd-Chiari syndrome is anticoagulation (blood thinning).

*Ascites,* already discussed, may also reflect liver failure.

## Why Are Liver Blood Tests Abnormal?

*Abnormal liver function tests* may be found in 30 to 60 percent of patients with lupus at some point and cause no symptoms. Blood enzyme evaluations included in routine blood panels such as the AST (also called SGOT), ALT (also called SGPT), alkaline phosphatase, and GGT may be elevated from a variety of mechanisms.

First of all, nearly all NSAIDs as well as acetominophen (Tylenol) can elevate these enzymes, and lupus patients—for unclear reasons—appear to be particularly susceptible to this. These abnormalities are usually of little consequence and generally represent false alarms, unless they are greater than three times normal. Also, active lupus can elevate these enzymes. Most nonsteroidals can be stopped for a week or two and the enzymes rechecked. If they remain increased, the possibilities for this elevation include hepatitis, infection, biliary disease, cancer, pancreatitis, alcoholism, or active lupus.

## What Is Autoimmune (Lupoid) Hepatitis?

Amanda did not feel right. She complained of vague right-sided upper abdominal discomfort for several months before seeing her doctor. Her examination demonstrated tenderness in this area, a low-grade fever, and distension. Blood tests revealed sky-high liver function studies. She was referred to a gastroenterologist, who embarked upon a workup that included a liver biopsy. It demonstrated chronic active hepatitis. Since Amanda did not have a history of viral hepatitis or alcohol abuse and since her hepatitis virus, cytomegalovirus, toxoplasmosis, and other tests were negative, autoantibodies were examined. These were consistent with autoimmune hepatitis. Amanda responded to steroids at first, but after 3 years began developing abdominal swelling, rectal bleeding, and mental slowing. She was switched to alpha-interferon along with prednisone and Imuran, to which Amanda responded. A special diet was strictly enforced and mild water pills were prescribed. This prolonged matters for another year, until it was evident that Amanda was experiencing liver failure. At that point she underwent a liver transplant and is doing fine 3 years later on prednisone, Imuran, and cyclosporine A.

*Lupoid hepatitis* is a complicated and controversial entity. Described in 1955 and coined by Ian Mackay in 1956, lupoid hepatitis has undergone many changes in definition. The overwhelming majority of patients who were told they had lupoid hepatitis between 1955 and 1975 would not fulfill current criteria. Initially thought of as the presence of chronic active hepatitis (''hepatitis'' means inflammation of the liver) with lupus erythematosus cells, the term *autoimmune hepatitis* seemed more appropriate, since few of these patients had typical clinical lupus. Patients with autoimmune hepatitis first notice right upper abdominal pain along with malaise, nausea, aching, and low-grade fevers. Loss of appetite, light-colored stools, and dark urine may also be present.

The development of diagnostic tests to detect hepatitis A, B, and more recently C has changed our concepts of autoimmune hepatitis. The current working definition of autoimmune hepatitis is (1) liver disease consistent with chronic active hepatitis; (2) absence of evidence for active hepatitis virus A, B, or C infection; and (3) a positive ANA or other autoantibodies associated with the syndrome.

Even using these criteria, only 10 percent of patients at the Mayo Clinic with autoimmune hepatitis fulfilled the American College of Rheumatology (ACR) criteria for lupus (Chapter 2). Many of the physical findings we associate with SLE (rashes, other organ involvement) are usually absent. Since lupus patients have compromised immune systems and can develop a viral hepatitis, take medications that can affect liver function, and some abuse alcohol just as nonlupus patients do (which can lead to a chronic active hepatitis), diagnosing autoimmune hepatitis can be tricky. About 30 to 60 percent of those with autoimmune hepatitis also have antibodies to smooth muscle or mitochondria (AMA or SMA).

### How Is Autoimmune Hepatitis Treated?

Why is it so important to split semantic terms to come up with the correct type of hepatitis? Because autoimmune hepatitis, if untreated, can be fatal within 5 years. However, it can respond to prednisone, steroids, alpha-interferon, and azathioprine. Two of our patients have undergone successful liver transplants, and we have come a long way in our understanding of liver disease in lupus and autoimmune hepatitis. But it still takes a great deal of skill to sort out what is an autoimmune disease and what is viral, and thus what medication is appropriate in each case.

## Summing Up

Difficulty in swallowing must be taken seriously. Heartburn and acid indigestion can be brought on by medication, stress, or active lupus. Nonspecific bowel symptoms are common and can be managed symptomatically unless a fever, localized tenderness, swollen abdomen, or bloody stools are present. These manifestations warrant prompt medical attention. The liver can be involved in lupus because of reactions to medication, antiphospholipid antibodies, or as a complication of infection. Vasculitis in the liver is rare and autoimmune hepatitis should be carefully looked for and ruled out.

# 19

# *Lupus in the Kidney and Urinary Tract*

The kidneys are critical to long-term health, and we have come a long way in understanding how they play a central role in many diseases. Unfortunately, in up to 40 percent of lupus patients, the disease can affect how the kidneys function. While some do not require treatment, most forms of kidney disease, termed *lupus nephritis,* warrant aggressive management. After discussing the functional anatomy of the kidney, the symptoms and signs of lupus nephritis are reviewed, followed by a classification of kidney disease and an overview of its therapy.

## HOW DOES THE KIDNEY WORK?

Think of the kidneys as the body's waste treatment plant. The body's wastes are filtered by the kidneys and then excreted in urine. The two kidneys are bean-shaped organs tucked neatly behind the abdominal cavity, level with the upper lumbar spine. Blood flows to the *nephron,* the functional unit of the kidney, and each kidney's approximately 1.2 million nephrons are divided into two parts. The first part is the *glomerulus,* a series of small, circular corpuscles through which materials are filtered and most reabsorbed back into the blood. The filtrated blood drains into the second part of nephron, the *tubule,* which both reabsorbs and secretes electrolytes (e.g., calcium, phosphorus, bicarbonate, sodium, potassium, chloride, and magnesium), glucose, and amino acids. These tubules form into the *ureter,* which connects to the *bladder*—the collecting area of materials destined for excretion. The kidney is principally responsible for maintaining the volume and composition of body fluids, excreting metabolic waste products, detoxifying and eliminating toxins, and regulating the hormones that control blood volume and blood pressure. Additionally, it is responsible for helping control the production of red blood cells and cell growth factors.

Lupus primarily affects the glomerulus and produces a condition known as *glomerulonephritis*. Figure 11 shows the functional anatomy of the kidney.

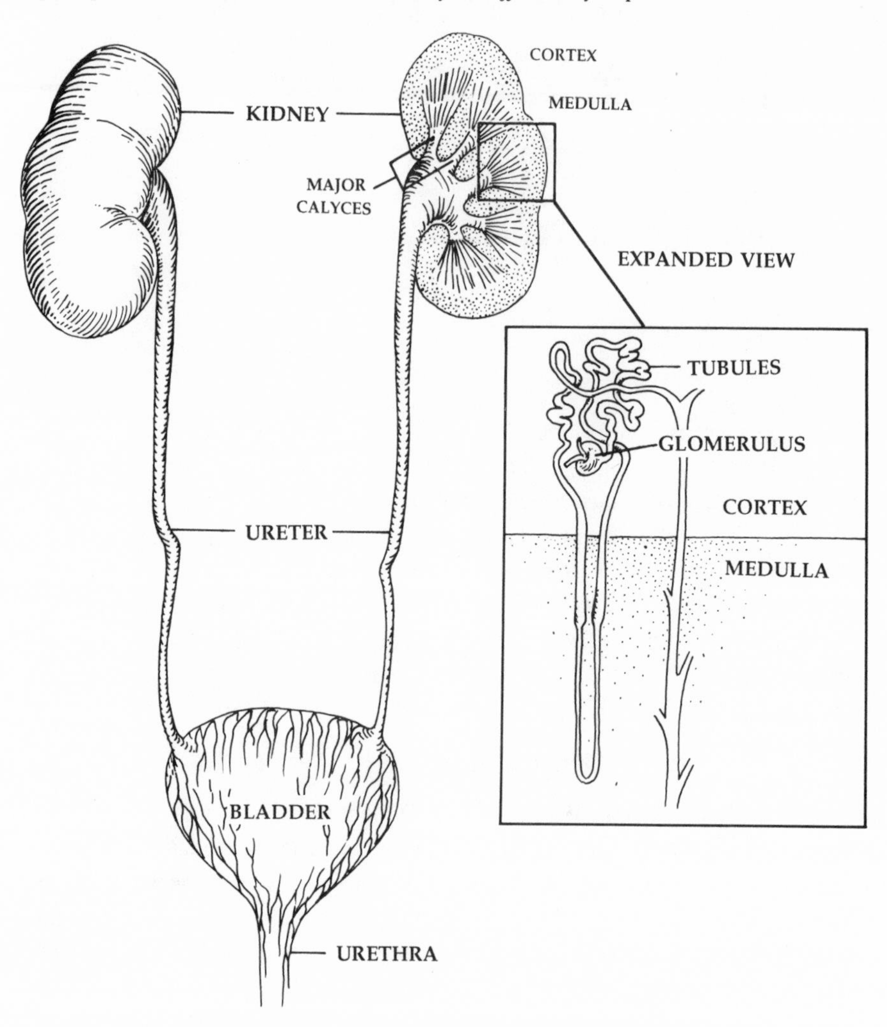


**Fig. 11** *The Female Kidney and Urogenital System*

## HOW CAN YOU TELL WHEN THE KIDNEY IS INVOLVED?

Lupus patients don't say, ''Doc, my kidney hurts!'' Pain in the kidney area would be felt if the patient had developed a kidney stone or severe kidney infection or had suffered a muscular spasm of the lumbar spine. None of these ailments have anything to do with lupus. In fact, most patients with kidney involvement have no specific complaints that can be immediately traced to the kidney. There are two circumstances that initiate awareness of a kidney problem in SLE patients: when they become *nephrotic* or *uremic*. In nephrotic syndrome, the kidney spills large amounts of protein due to a filtering defect and serum albumin levels become very low. When serum albumin levels drop below 2.8 grams per deciliter (g/dL) or per 100 milliliters and 24-hour urine protein measurements rise above 3.5 grams, swelling is apparent in the ankles and abdomen. Lower amounts of protein loss may result in mild ankle swelling. The patient may complain of a general sense of bloating and discomfort. Pleural and pericardial effusions may be noted, and shortness of breath or chest pains are occasionally present. In uremia, a failing kidney inadequately filters wastes and toxic materials accumulate, which can damage other tissues. When a kidney is functioning at a level below 10 percent of normal, the patient will probably complain of fatigue, look pale, and emit a distinct odor. Patients suffering from advanced uremia must go on dialysis in order to live.

Since patients with lupus nephritis usually have few obvious symptoms, how can we tell whether the kidney is involved? We do it through blood and urine testing. Elevations in the serum *blood urea nitrogen* (*BUN*) or *creatinine* usually reflect an abnormality in kidney function and indicate renal involvement. A *urinalysis* is the most accurate assessment for the presence of lupus nephritis. Patients with SLE have cellular debris called *casts* visible in urine when viewed under the microscope. Many different types of casts can be found, including hyaline, granular, or red cell casts. Most nephritis patients have *hematuria,* or at least small amounts of microscopic blood in their urine. The most specific correlate of lupus nephritis is *proteinuria,* or protein in the urine. The presence of protein can be determined quickly and painlessly with a urine test in the doctor's office. If proteinuria is found, most physicians attempt to quantitate it by having the patient collect the urine for 24 hours; then the amount of protein in the sample is measured. Expect the *24-hour urine protein* to exceed 300 milligrams when lupus is active in the kidney. Nephrotic patients have protein levels between 3500 and 25,000 milligrams. As part of a 24-hour urine collection, the *creatinine clearance* can be calculated. In theory, creatinine clearance estimates kidney function more reliably than the serum creatinine, but in my experience it is much less helpful and varies by up to 50 percent in a 24-hour period.

## WHAT ARE THE TYPES OF LUPUS GLOMERULONEPHRITIS?

When lupus affects the kidney, it makes sense for a rheumatologist like me to seek consultation from a kidney specialist (nephrologist) to obtain a renal biopsy. A renal biopsy is recommended if there is abnormal urine sediment (e.g., casts, hematuria) and more than 500 milligrams of protein in a 24-hour urine specimen. Renal biopsies have been performed for three reasons: (1) to confirm the diagnosis of lupus nephritis as opposed to another disease; (2) to determine if the kidney tissue is inflamed, scarred, or both; and (3) to evaluate treatment. Treating a patient with an elevated creatinine who has significant scarring but no inflammation with anti-inflammatory medication for the kidney is not advisable. There is no reason for using potentially toxic medicines without cause.

Tissue obtained at biopsy is examined by three methods. First, it is stained to look for structural abnormalities and viewed under a standard microscope. Second, the sample is evaluated for antibodies to the gamma globulins IgG, IgA, and IgM under an immunofluorescent microscope. Finally, the tissue specimen is examined with an electron microscope to search for ''electron-dense deposits,'' which are immune complexes that interfere with the kidney's ability to filter materials properly. Most nephrologists obtain enough material to give the pathologist an opportunity to use all three methods. Occasionally, only a small amount of tissue is available, which permits a more limited evaluation.

These methods allow kidney tissue to be classified according to the *World Health Organization*'s system of six different patterns. In *Class I,* the light microscopy is normal and electron microscopy shows minimal abnormalities. No treatment is indicated. *Class II* disease is termed *mesangial* and reflects mild kidney involvement. Low doses of steroids are sometimes given. *Class III* is called *focal proliferative* nephritis, while *Class IV* is *diffuse proliferative*. Proliferative disease, both types being extensions of the same process, is a serious complication and will usually lead to *end-stage renal disease (ESRD)* necessitating dialysis if it is not treated. *Class V* glomerulonephritis is known as *membranous* and is characterized by a high incidence of nephrosis with a tendency towards a slow, indolent, progressive course ending with renal failure if not treated. *Class VI* nephritis is known as *glomerulosclerosis* and represents a scarred down, end-stage kidney with irreversible disease.

A doctor can perform a kidney biopsy in a hospital at the bedside or in a radiology suite with ultrasonic guidance. An ultrasound examination should confirm the location of the kidneys and that the patient has two of them. The risk of bleeding from the biopsy in experienced hands is small—1 in 100 patients needs a blood transfusion and 1 in 1000 will have a serious complication. Most patients are able to go home within 36 hours of the biopsy without any work or activity restrictions.

Occasionally, renal biopsies are ill advised or require special preparation. These circumstances include patients who weigh over 250 pounds (they require an open biopsy at surgery), patients who are very anemic and/or who cannot risk blood transfusions (e.g., Jehovah's Witnesses), or individuals who have unique clotting problems that predispose them to unusual bleeding risks or require continuous anticoagulation.

## HOW DO WE MONITOR KIDNEY DISEASE?

There is no one test that best assesses lupus in the kidneys. The BUN and creatinine tell doctors how well the kidney is functioning. High blood pressure signals that the kidney is under stress, and persistent elevations are associated with the development of kidney failure. Low serum albumins and high 24-hour urine proteins tell physicians how much leakage there is. Low C3 complement levels and high anti-DNAs in the blood indicate active lupus inflammation. Microscopic evaluation of a freshly voided urine specimen can roughly suggest how active the nephritis is. I follow nephritis patients by taking their weight (to measure fluid retention) and blood pressure. Blood and urine are obtained and a 24-hour urine collection is measured if indicated. These evaluations will reveal a general pattern of improvement, stabilization, or worsening and will suggest alterations of therapy if needed.

## ARE THERE OTHER FORMS OF LUPUS KIDNEY DISEASE BESIDES GLOMERULONEPHRITIS?

Although they rarely cause symptoms, many *commonly used drugs can affect renal function,* the most common of which are the NSAIDs. Some of them, particularly indomethacin, can raise serum creatinine as part of their action against a chemical called prostaglandin. A normal serum creatinine is up to about 1.5 mg/dL. If the creatinine is slightly above this level, nonsteroidals should probably be used only for a few days and the blood should be carefully monitored. Circumstances when this might arise would be an acute attack of gout or bursitis. Since creatinines are measured in terms of a logarithmic function rather than an arithmetic one, a serum creatinine of 2 indicates only 50 percent kidney function, 3 means 30 percent kidney function, and 4 signifies 20 percent kidney function. (Remember logarithms from high school? A rise in creatinine from 2 to 3 indicates a 10-fold change!) Patients with a serum creatinine above 6 mg/dL usually require dialysis. The use of NSAIDs is not advised if the creatinine is above 2.

Other drugs can induce an *interstitial nephritis,* which inflames and then scars the connective tissue of the glomerulus. Certain antibiotics and anti-inflammatory drugs are included in this group. Sjögern's syndrome is associated with *tubular dysfunction,* which leads to electrolyte abnormalities. High blood pressure also damages kidney tissue and is capable of inducing renal failure regardless of SLE activity. Patients with antiphospholipid antibodies are predisposed toward developing *renal vein thrombosis,* especially if they have membranous nephritis. This condition can result in acute renal failure, flank (side of the back) pain, and fever. It is managed with anticoagulation and steroids if the lupus is active.

## WHAT IS THE NATURAL COURSE OF LUPUS NEPHRITIS AND HOW DO WE TREAT IT?

When Dr. Cohn told Gillian she had lupus in her kidney, it was not an easy thing to say to a promising concert pianist. At the time, her ankles were so swollen that Gillian couldn't wear any shoes other than an old pair of sneakers. Her anti-DNA levels were high and serum complement levels low. Dr. Cohn arranged for a kidney biopsy, which showed class IV (diffuse proliferative) disease with a lot of activity and little scarring. He told Gillian that she had a reversible lesion and needed chemotherapy as well as steroids. She was started on 60 mg of prednisone a day and received six monthly doses of intravenous Cytoxan at the hospital infusion center as an outpatient. During this time, Gillian gained 40 pounds, became moody and irritable, and could not concentrate on her practicing. After 6 months, she began to improve. Her creatinine was still normal at 1.4 and her 24-hour urine protein had decreased from 7 grams to 3. At this point, Dr. Cohn extended the Cytoxan treatments to every 3 months and tapered her prednisone to 10 mg a day. Gillian did well for a year, but her blood pressure began to go up, as did her cholesterol and blood sugars. She was started on blood pressure medicine and put on a strict low-fat, low-carbohydrate diet. After 2 years, the Cytoxan was stopped. Gillian was maintained on 10 mg of prednisone a day to suppress her renal disease and her mild lupus symptoms outside the kidney. Five years later, her creatinine was 2.6, along with a 24-hour urine protein of 1.3 grams. Dr. Cohn arranged for a second biopsy that showed little lupus nephritis activity but a lot of damage from hypertension and scarring. Her medicine was kept the same. Being on the road a lot, Gillian tried the best she could to keep to her diet, but it was not easy. After 10 years of nephritis, Gillian's creatinine finally crept up near 7 and she was placed on hemodialysis. Fortunately, her brother was able to donate a kidney and she underwent a successful transplant. Fifteen years later, her creatinine is 1.0 and she is on 10 mg of prednisone a day, Cyclosporin A to prevent transplant rejection, a diabetes medicine, a cholesterol medicine, and a blood pressure medicine. At age 30, Gillian teaches music at a junior college.

Patients with class I and II biopsy patterns have an excellent outcome. Class V is relatively resistant to therapy even though most rheumatologists try a course of corticosteroids with or without *immunosuppressive regimens*. Some of the drugs that fall into this category include azathioprine, cyclophosphamide, chlorambucil, or cyclosporin A, all of which are discussed in Chapters 27 and 28. Class VI usually leads to end-stage renal disease within months and no specific therapy is indicated. Class III or IV proliferative nephritis is reversible some of the time with aggressive therapy. High doses of corticosteroids with or without immunosuppressives (particularly intravenous cyclophosphamide, or Cytoxan) can prolong kidney function and prevent the need for dialysis over a 10-year period in half the patients.

Is there anything you can do to prevent the effects of kidney disease? High blood pressure should be managed aggressively, since it accelerates functional kidney impairment. Stress reduction helps lower blood pressure. Patients with renal disease should restrict their salt intake to no more than 3 grams a day; when renal function is 50 percent or less, normal protein intake should also be restricted. Diuretics help remove fluid in lupus patients and make them feel more comfortable, but they must be used cautiously, since they alter electrolyte balance. Potassium supplementation is given to patients with normal kidney function on most diuretics. However, potassium intake is restricted in patients with markedly impaired renal function.

## HOW IS THE URINARY TRACT AFFECTED BY LUPUS?

The ureter is not involved in lupus and the bladder is a rare target of the disease. But a condition known as *lupus cystitis* is observed in 1 to 5 percent of those with lupus. Manifested by inflammation of the lining of the bladder with blood in the urine, cystitis can be diagnosed by an office procedure known as cystoscopy. Lupus cystitis frequently correlates with gastrointestinal malabsorption. It is treated with antibiotics and anti-inflammatory lupus medications. Occasionally, a drug called DMSO (dimethylsulfoxide) is administered directly into the bladder through a cystoscope.

Young women are especially prone to develop urinary tract infections, and young women with lupus are particularly vulnerable to infections in general. The drugs of choice for most urinary tract infections are sulfa antibiotics, but these are often poorly tolerated by SLE patients (Chapter 9). A common problem I encounter takes place when gynecologists, urologists, or family practitioners prescribe sulfa antibiotics without realizing that lupus patients frequently develop flareups of SLE when they are given these drugs. Many antibiotic alternatives are available in these circumstances. Always tell physicians who are treating you that they are dealing with a lupus patient.

## THE BOTTOM LINE

Lupus nephritis is very tricky to treat, since it produces few symptoms or signs. Despite our best efforts, patients may still evolve kidney failure. First, a doctor must confirm that it is indeed lupus that is affecting the kidney. Second, renal function and inflammation must be measured. On the basis of blood tests, urine testing, and biopsy material (if available), a careful treatment plan is formulated. Because many of the drugs used to prevent or retard this disease are quite potent in their own right, a careful balance is important.

The reader is referred to other parts of this book for details regarding dialysis and transplantation (Chapter 28); cyclophosphamide, chlorambucil, nitrogen mustard, azathioprine, pulse steroids, steroids, and apheresis management (Chapter 27 and 28); and pregnancy in the patient with lupus nephritis (Chapter 30).

# 20

# *The Blood and Lymphatic Systems*

Before rheumatology was recognized as a subspecialty of internal medicine in 1972, lupus patients were treated primarily by another group of subspecialists—hematologists. Hematology is the study of diseases of the lymph glands and blood components, all of which play a key role in the well-being of individuals with SLE. This chapter looks at why many lupus patients have blood and lymphatic abnormalities. Although rheumatologists supervise the overall management of SLE, hematologists still play an important role in managing the blood abnormalities seen in lupus.

## WHAT'S WRONG WITH THE PATIENT'S BLOOD?

Whole blood is divided into three major components: red blood cells, white blood cells, and platelets. The function of each of these is reviewed in Chapter 5. In SLE, any or all of these components may be *quantitatively* (or numerically) high or low, or they may be malfunctioning *qualitatively*. A simple tube of blood measures quantitative values; qualitative tests also require blood samples but are more difficult to perform.

## COULD THE PROBLEM BE ANEMIA?

Are you tired? Do you feel weak? Do you look pale? If any of these answers are yes, you could be anemic. About 80 percent of SLE patients are anemic during the course of their disease. Anemia, or a low red blood cell count, is defined as a *hemoglobin* count of less than 12 grams per deciliter (g/dL) or a *hematocrit* of less than 36. The hematocrit is the percentage of red cells per 100 ml of blood and it is usually about three times the hemoglobin. A normal hemoglobin ranges from 12 to 16 g/dL and indicates the amount of a certain protein in red blood cells. Anemia in lupus is divided into two general categories: nonimmunologic and immunologic. The major symptom of anemia is fatigue, which is usually evident when the hemoglobin drops into the 10 g/dL range. Further decreases can make a person appear pale and feel weak.

## Nonimmunologic Anemias

Red blood cells are made by the bone marrow and released into the circulation. When people develop chronic inflammatory disorders, the stimulus to make red blood cells decreases. These individuals develop what is called an *anemia of chronic disease.*

> A week after Sylvia had her annual gynecologic evaluation, her doctor called her to say that she was anemic. Her hemoglobin was 9.7, and it had been 13 a year before. Sylvia's lupus had been under excellent control with three aspirins four times a day, and even though she felt tired, Sylvia had only minimal aching. Her periods had always been heavy, but not unusually so. She consulted her internist, who performed a thorough evaluation. He found that Sylvia's anemia stemmed from several causes: She was iron-deficient due to her heavy periods, had an anemia of chronic disease, and was also having a problem with the aspirin. An endoscopy of her stomach showed evidence of gastritis from taking aspirin. Her doctor started her on iron and ranitidine (an ulcer medicine, with the brand name Zantac) and stopped aspirin for 6 weeks. When her hemoglobin rose to 12 and another endoscopy showed the ulcer healing, an aspirin derivative (Disalcid) that rarely causes ulcers was started and she was kept on the iron.

Lupus patients can be anemic for any of the reasons that makes otherwise healthy people become anemic. However, several of these causes are more prominent in SLE, especially when superimposed upon a preexisting anemia or chronic disease. Among young women a common cause is heavy menstrual bleeding and resultant *iron-deficiency anemia.* Another cause of this type of anemia is the administration of nonsteroidal anti-inflammatory drugs (NSAIDs) such as ibuprofen or aspirin, which can irritate the stomach lining, inducing secondary blood loss from an erosive gastritis. Practitioners often check the stools of their lupus patients for blood on a periodic basis if these patients are taking nonsteroidals. If continued NSAID administration is necessary, patients with these problems are given iron supplements and gastric agents that coat the stomach. However, doctors often discontinue the NSAIDs.

A hormone made by the kidney, called *erythropoietin* (EPO), stimulates the bone marrow to make red blood cells. *Chronic renal disease* is associated with decreased EPO levels, which can lead to anemia. The availability of EPO injections has greatly improved these patients' quality of life over the last decade. There are many other causes of anemia in lupus, but they occur in the same proportion as in the general population.

## Immunologic Anemias

Mary was a healthy college student until she suddenly became profoundly fatigued and started noticing a yellowish cast to her skin. At a victory celebration, after her team had won its Saturday football game, she passed out. Paramedics were called and took her to a hospital, where her hemoglobin was found to be 6. An internist then undertook an anemia workup. It turned out that she had SLE with autoimmune hemolytic anemia. Even though her bone marrow was churning out red blood cells, they were being destroyed within days upon release into the circulation. A hematologist was consulted and started her on 60 milligrams of prednisone daily. Twelve weeks later, she had a hemoglobin of only 9 and azathioprine was added to the regimen.

Up to 10 percent of patients with lupus develop *autoimmune hemolytic anemia* (*AIHA*); most of these patients complain of weakness, dizziness, and fevers. They may appear jaundiced (have a yellowish complexion) as a result of the rapid destruction of red blood cells. Antibodies to the surface of red blood cells are responsible for AIHA. A red blood cell normally lives for 120 days, but in AIHA it is destroyed by antibodies much earlier. The bone marrow is then stimulated to make more red blood cells, which can sometimes compensate for this early destruction, as evidenced by an elevation in the reticulocyte count (which measures rate of formation of red blood cells), but most of the time it is not enough.

Specialized blood testing can help to diagnose AIHA, and the deformed cells can be seen by looking at a blood smear under the microscope. Suspicions that AIHA is present can be confirmed by blood tests (such as an elevated reticulocyte count, serum LDH, decreased serum haptoglobin, or the presence of an antibody to red blood cells which is measured using the Coombs' antibody test).

This potentially serious complication of SLE responds only temporarily to transfusions and often calls for a prolonged course of high-dose corticosteroids. Poorly responsive AIHA patients have their steroids supplemented with cyclophosphamide, azathioprine, or danazol. Occasionally, a splenectomy or removal of the spleen is necessary, since this organ removes partially damaged red cells, thus contributing to anemia.

In rare cases, circulating plasma factors tell the bone marrow to turn off the production of red blood cells. The end result of this is a slow-down or shut-down of the bone marrow, termed *hypoplasia* or *aplasia*. This usually reflects active disease, and corticosteroids or cytotoxic drugs are necessary (see Chapter 27). A careful drug history should be taken, since many prescription drugs (e.g., sulfa antibiotics) can occasionally cause aplasia.

### Can the Patient Have a Blood Transfusion?

Several studies have shown that lupus patients have the same blood types and distribution of blood types as the general population. However, patients with immune-mediated anemias break down transfused cells more rapidly than otherwise healthy individuals. One study suggested that up to 16 percent of lupus patients with immune anemias experience a mild allergic-type reaction when their blood is transfused. This reaction results from red cell antibodies. There is usually no problem transfusing lupus patients without immune anemias. An extra does of corticosteriods or an antihistamine such as Benadryl can be ordered immediately prior to the transfusion to minimize any reactions. With current screening methods in the United States, the risk of AIDS or other transmissible viruses is on the order of one in tens of thousands, and the use of directed donors (e.g., having friends or relatives donate blood for you) further decreases this risk. On the other hand, it is not a good idea for lupus patients to donate their blood, since it contains too many antibodies.

## HOW IMPORTANT ARE WHITE BLOOD CELLS?

Half of all lupus patients develop low white blood cell counts during the course of their disease. White blood cells, or leukocytes, constitute the body's defense mechanism. They are responsible for immunologic memory (lymphocytes), bacterial killing (neutrophils), and allergic reactions (eosinophils). Chapter 5 reviews the functions of the five types of white blood cells. White blood cell counts are elevated by active infections, corticosteroid therapy, and sometimes by active SLE. Low white blood cell counts derive from viral infections or active lupus or are the consequence of chemotherapy for lupus.

The usual reason for low *lymphocyte* levels is the presence of antilymphocyte antibodies, which destroy lymphocytes. *Neutrophil* counts are usually normal in SLE unless suppressed by chemotherapy. Yet though their levels are normal, there are apparently qualitative defects in neutrophil function which are manifest by a decreased ability to kill bacteria. This helps explain why lupus patients are more susceptible to infection. *Eosinophils* are elevated in 3 to 10 percent of lupus patients, and lupus patients have an increased incidence of allergies compared with the general population (Chapter 29).

## CLOTTING AND BLEEDING PROBLEMS

Platelets are the blood components responsible for clotting blood. Increased numbers of platelets reflect acute inflammation and are observed in a minority of patients with active lupus. Low platelet counts are associated with bleeding disorders. The lupus anticoagulant usually promotes clotting and can damage the body due to functional abnormalities of the platelets (see Chapter 21).

When Celeste was in the sixth grade, she noticed red spots on her legs. She told her mother, who took her to see their pediatrician. Dr. Hawkins obtained a blood count which indicated that Celeste had 30,000 platelets per cubic milliliter. Celeste was started on prednisone, after which her platelet count became normal. She was off all medication within a year. Six years later, after a bad flu, the spots reappeared and her platelet count dropped to 20,000 cubic milliliters. Her internist also gave her an ANA test that proved to be positive. Even though she denied any of the signs or symptoms of lupus and enjoyed being out in the sun, Celeste was told she had lupus with idiopathic thrombocytopenic purpura. She was started on prednisone and her platelet counts normalized within a few weeks. Celeste now feels fine and is off all medication.

Decreased platelet counts along with the presence of platelet antibodies is called *idiopathic thrombocytopenic purpura* (*ITP*). It is usually seen in children and young women, most of whom have no symptoms other than that they bruise easily. Approximately 20 percent of patients with ITP also have a positive antinuclear antibody test, and 20 percent of these patients ultimately develop lupus. A normal platelet count is between 150,000 and 400,000 per cubic millimeter. One-sixth of all lupus patients run platelet counts below 100,000 during the course of their disease. Easy bruising is noted when the counts drop below 50,000, and counts less than 20,000 can be life-threatening in the sense that these individuals can suffer spontaneous internal bleeding. Anyone who notices numerous black-and-blue marks, excessive bleeding from the gums, very heavy periods, or little red spots (petechiae) on the skin should obtain a platelet count. Many ITP patients lack other signs and symptoms of SLE, and many are not even sun-sensitive.

I usually use a drug called danazol in low doses (Chapter 28) to treat platelet counts that fall between 60,000 and 100,000 per cubic milliliter and I add steroids only if lupus is active outside the platelet system or the counts drop further. This condition is usually fairly responsive to steroids, but occasionally I add azathioprine, vincristine, or cyclophosphamide (Chapter 27). Intravenous gamma globulin and platelet transfusions temporarily raise platelet counts, as does plasmapheresis. Removal of the spleen, the organ that traps platelets coated with antibodies, is usually curative. Most patients with ITP and lupus also have antiphospholipid antibodies (Chapter 21).

A rare and frightening complication of SLE is *thrombotic thrombocytopenic purpura* (*TTP*). Called a ''pentad'' because of its five signal markers of fever, hemolytic anemia, neurologic impairment, kidney failure, and low platelet counts, TTP can lead to multiple organ failure when disseminated clots form throughout the body. It can be brought on by a viral infection or other forms of sepsis and was fatal until recently. The critical lifesaving element is the treating doctor's awareness of this rare disorder as a possibility and the ability to diagnose it promptly. Plasmapheresis can cure the condition, which rarely comes back once it has been successfully treated.

Lupus patients also have *qualitative platelet defects.* Aspirin, platelet antibodies, and chronic renal failure can all alter functional blood clotting even when palatelet counts are in the normal range. Steroids and NSAIDs disrupt platelets and induce *purpura,* those black-and-blue marks on the skin that result partially from damage to fragile capillaries. This benign but annoying condition causes no symptoms and need not cause alarm if platelet counts are in the normal range. No treatment is necessary.

## DOES LUPUS CAUSE SWOLLEN GLANDS?

When the disease is active, the increased numbers of inflammatory cells and immune complexes can cause lymph glands (or nodes) to enlarge. Lymph glands, loosely arrayed in chains throughout the body, drain and filter particulate materials. Whereas arteries supply blood and nutrients to the body, veins return blood and nutrients to the heart and are helped along by lymph glands. Half of all lupus patients have enlarged lymph nodes that can be felt on a physical examination at some point during the course of the disease. On occasion, the nodes can be up to 1.5 inches in diameter. Infections also enlarge lymph glands, as can malignancies; before active lupus is treated, these possibilities must be ruled out. *Lymphadenopathy* (another name for swollen lymph glands) from SLE is treated with anti-inflammatory medication.

## HOW IS THE SPLEEN INVOLVED IN LUPUS?

In the left upper part of the abdomen lies a large, vascular lymphatic organ called the *spleen.* The spleen filters the blood and destroys and removes damaged red blood cells, white blood cells, and platelets. It can become larger when a greater number of cells than usual require removal. This process is part of what rheumatologists and immunologists call the *reticuloendothelial system* (RES).

Circulating immune complexes are also cleared by the spleen, and in SLE their clearance can be impaired. Impaired RES clearance increases the deposition of immune complexes in tissue, which, in turn, causes damage or inflammation. About 10 percent of lupus patients have enlarged spleens on physical examination. Abdominal computed tomography or ultrasound easily shows an enlarged spleen. The spleen rarely produces pain unless it is very large and the capsule is stretched. *The most common cause of left upper abdominal pain in SLE is pleurisy, since the lung and ribs overlie the spleen.*

On occasion, the spleen is removed to treat AIHA or ITP. Even though we can live a normal life expectancy without a spleen, some individuals become vulnerable to numerous infectious agents, especially a form of bacterium that leads to pneumococcal pneunomia. Every patient who has had the spleen removed (splenectomy) should be vaccinated against this bacterium and watched closely for infection.

## WHAT DOES THE THYMUS DO?

As discussed earlier, the thymus is a lymph organ at the base of the neck; it is responsible for establishing a system of immune surveillance. In adult life, the gland atrophies and is barely recognizable. The thymus in lupus patients does not look any different than it does in a healthy person. Removing the gland, with a *thymectomy,* usually has no effect upon the disease.

### Summing Up

Anemia, the most common blood abnormality in lupus patients, can cause fatigue and pallor. The anemia can result from a nonimmunological cause—iron deficiency or chronic disease—or it can result directly from immunological conditions caused by lupus. Immune-mediated anemias are serious and potentially life-threatening. They often mandate high doses of steroids and other immunosuppressive therapies.

Low white blood cell counts are also commonly observed and result from antilymphocyte antibodies. White blood cells (called neutrophils) do not kill bacteria as well as they should in SLE, and this increases the risk of infection.

Antibodies to platelets lower platelet counts, especially in patients with the lupus anticoagulant. Very low platelet counts can result in serious internal bleeding and must be managed with steroids and other immunosuppressive therapies.

Lymph glands swell with active lupus, but this can be treated. The spleen also enlarges when its filtering capacities are overwhelmed, but treatment can resolve this problem as well.

# 21

# *Why Do Blood Clots Develop?*

As many as one-third of all deaths due to complications from lupus arise from blood-clotting abnormalities. The saga of how we came to see that patients with SLE were especially susceptible to blood clots is one of the more interesting and convoluted tales in rheumatology. After 40 years of struggling with the problem, rheumatologists finally realized that the solution lay in tying the knot right under our nose. Fortunately, rapid developments in this area over the last decade should greatly decrease complication and mortality rates in this group at risk.

## LUPUS AND FALSE-POSITIVE SYPHILIS TESTS

In 1940, Dr. Harry Keil and his colleagues at Johns Hopkins linked 10 women together with a very peculiar finding. They all had lupus and they all tested positive for syphilis, but they did not have the venereal disease. Many of these tests were performed as part of routine premarital exams. I still remember several patients relating to me that in the 1940s and 1950s they were told they had syphilis, even though they were virgins. Engagements were broken and misunderstandings abounded. Further studies showed that up to 20 percent of all lupus patients had a false-positive "Wassermann" test, as the syphilis test was called in those days. However, these patients with lupus had no unique or specific clinical features that differentiated them from others with the disease.

The Wassermann test relied upon *reagin,* an antibody found in syphilis patients. Further work showed that the antigen to which this antibody reacted was *cardiolipin,* a phosphorus-fat component of cell membranes called *phospholipid.*

## WHAT IS THE LUPUS ANTICOAGULANT?

In 1948, another Johns Hopkins team led by Dr. C. Lockard Conley and his colleagues reported that an antibody found in the blood of lupus patients prolonged phospholipid-dependent clotting tests. In time, it was called the *lupus anticoagulant.* This term has turned out to be a misnomer, since—except in unusual circumstances—it is associated with the formation of blood clots rather than increased bleeding. For years, no investigator linked patients with false-positive syphilis tests to those who had the lupus anticoagulant. It did not seem important, since these findings were considered laboratory curiosities and lupus patients were strange in any case. Little if any clinical relevance was attached to these oddities for some 35 years.

## PHOSPHOLIPID ANTIBODIES COME OF AGE

In the early 1980s, a team of English investigators headed by Nigel Harris began looking in earnest at antibodies to the troublesome phospholipid antigens. Using newly available immunologic techniques, they identified several antiphospholipid antibodies that had important clinical implications. One of these antibodies, the *anticardiolipin antibody,* was first correlated with an increased risk of *thromboses,* or blood clots. As the testing process was further refined, a myriad of clinical associations were confirmed. *Approximately one-third of all lupus patients possess antiphospholipid antibodies, and one-third of these patients have complications as a result of this antibody.* Since several antiphospholipid antibodies are associated with blood clots, what was originally called the anticardiolipin syndrome is now termed the *antiphospholipid syndrome.*

## WHAT IS THE ANTIPHOSPHOLIPID SYNDROME?

As mentioned above, many patients have antiphospholipid antibodies but only a small proportion of these individuals have problems resulting from them. For example, 10 to 30 percent of all patients with rheumatoid arthritis, scleroderma, and other forms of vasculitis have antiphospholipid antibodies, even though clotting complications are extremely unusual. Further, many infectious diseases, particularly AIDS, are associated with these antibodies and never cause clotting problems. The reasons for this are twofold. First, lupus patients are uniquely susceptible to the antiphospholipid syndrome because they have an additional protein that makes the antiphospholipid antibody promote clots. Also, antiphospholipid antibodies are directed against different *isotypes,* or types of immunoglobulins, against which the antibody is directed (Chapter 11). IgG anticardiolipin antibody, for example, is much more likely to lead to blood clots than IgM or IgA anticardiolipin antibody. Higher amounts of antibody also increase the risk of clots.

The term *antiphospholipid syndrome* is applied to a group of clinical complications that result from antiphospholipid antibodies. The overwhelming majority of patients with this syndrome have lupus, but a very small percentage of otherwise healthy people have the antiphospholipid syndrome.

## WHAT IS THE IMPORTANCE OF ANTIPHOSPHOLIPID ANTIBODIES?

Quite simply, antiphospholipid antibodies and the lupus anticoagulant can cause blood clots and blood clots are potentially serious. These clots can form anywhere in the body, especially in arteries or veins. If they appear in the brain, they can produce a stroke (Chapter 18). In Libman-Sacks endocarditis, the heart valves

**Table 11.** *Complications Caused by Antiphospholipid Antibodies in Lupus*

| |
|---|
| Obstetric |
| Fetal loss/miscarriages |
| Hematologic |
| Arterial and venous clots (thromboses) |
| Low platelet counts (autoimmune thrombocytopenia) |
| Anemia (autoimmune hemolytic anemia) |
| Neurologic |
| Strokes |
| Migraines |
| Transient ischemic attacks (TIAs, or stroke warnings) |
| Cardiologic |
| Libman-Sacks endocarditis |
| Pulmonary |
| Pulmonary emboli |
| Pulmonary hypertension |
| Joints |
| Avascular necrosis |
| Dermatologic |
| Livedo reticularis |
| Ulcers and gangrene |

can become a source for infection and can produce emboli (traveling blood clots) to the brain, which lead to strokes (Chapter 14). In the body's vascular system, phlebitis (inflammation of a vein) from a clot is not uncommon, especially in the calves of the legs. Sometimes leg clots can travel to the lungs and produce pulmonary emboli (Chapter 14). Multiple pulmonary emboli may lead to pulmonary hypertension.

In our blood, antibodies to platelets or red blood cells can be closely associated with antiphospholipid antibodies. Several serious conditions are associated with these antibodies including autoimmune hemolytic anemia and thrombocytopenia. Pregnant women with antiphospholipid antibodies can miscarry and must be closely monitored (Chapter 30). Other conditions are also found in this syndrome. Among them are avascular necrosis (or dead bone) (Chapter 13), and livedo reticularis of the skin (Chapter 12).

The risks of clotting are not necessarily related to disease activity and can become troublesome when lupus is in remission. Table 11 summarizes these findings and shows how antiphospholipid antibodies can be complicating factors in lupus that relate to many different parts of the body.

## WHY DOES ABNORMAL CLOTTING DEVELOP?

We still don't know why antiphospholipid antibodies predispose patients to blood clots. Several theories that are difficult to test have been put forward. Current thinking suggests that these antibodies bind to platelets and activate them. This combination increases the risk of forming clots. Or possibly antiphospholipid antibodies could bind to *endothelial cells,* the cells that line blood vessels, or inhibit the release of certain chemicals that dilate blood vessels. Patients with lupus are prone to develop an acquired deficiency of several proteins important in clotting. These include protein C, protein S, and antithrombin 3. A lack of any of these proteins induces what physicians call a *hypercoagulable state,* or a milieu where clotting risks are high.

A small percentage of patients with antiphospholipid antibodies are more likely to experience bleeding then clotting. This group has either very low platelet counts—less than 30,000 per cubic millimeter (30,000/$mm^3$); normal is more than 150,000/$mm^3$—or lack a blood clotting factor called factor II (also known as prothrombin).

## WHAT IS THE RELATIONSHIP BETWEEN ANTIPHOSPHOLIPID ANTIBODIES, THE LUPUS ANTICOAGULANT, AND FALSE-POSITIVE TESTS FOR SYPHILIS?

Since many patients with antiphospholipid antibodies feel well until they develop a clot, it is frequently difficult to convey the complicated interactions between clotting factors and antibodies in a meaningful way. Figure 12 is designed to help the reader visualize these interrelationships.

The majority of patients with the lupus anticoagulant also have positive tests for antiphospholipid antibodies, and vice versa. As previously noted, many of these individuals also have a false-positive syphilis test. The prevalence of the lupus anticoagulant depends on how the testing is performed. A *partial thromboplastin time* (*PTT*) measures how long it takes to activate the body's intrinsic clotting cascade. Most lupus patients have normal PTTs when tested by conventional methods. However, the PTT can be modified by a variety of methods that reduce the amount of phospholipid in the test clotting mixture. This brings out evidence of an antibody to the *prothrombin activator complex,* or clotting factors X and V (ten and five), which is an antiphospholipid antibody. It is probably not anticardiolipin antibody but is closely related. That's why some patients with a positive anticardiolipin antibody can have a negative modified PTT and others with a positive lupus anticoagulant test have negative anticardiolipin antibody testing.

Some of the names for lupus anticoagulant tests include the Russell viper venom test, the RBNP, and the kaolin PTT. Ten percent of patients with SLE have an abnormal conventional PTT and 30 percent a prolonged modified PTT. Another 20 percent of lupus patients have a false-positive VDRL, which detects antiphospholipid antibodies that are not usually associated with clotting problems.

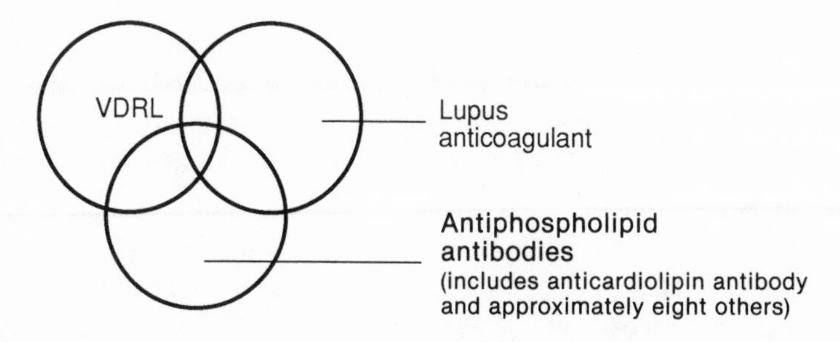


**Fig. 12.** *Relationships between Various Methods for Detecting Antiphospholipid Antibodies*

## WHAT TESTING SHOULD BE DONE TO SCREEN FOR CLOTTING RISKS?

Most of my new lupus patients are screened for the lupus anticoagulant and anticardiolipin antibody and have a syphilis serologic test performed. The cost of doing these three tests is less than $200, and they can be lifesaving. If these tests are negative or only borderline positive in patients with an abnormal clotting history, I measure protein C, protein S, and antithrombin 3 levels and look for the presence of other antiphospholipid antibodies for which testing is now commercially available. Anticardiolipin antibodies can appear and disappear as the disease waxes and wanes. Steroids can decrease anticardiolipin levels or make them disappear altogether. Some lupus patients have antiphospholipid antibodies present only when they are pregnant. Other less common deficiencies of additional clotting factors occasionally induce clots, and a hematology consultant may wish to test for these.

## HOW SHOULD ANTIPHOSPHOLIPID ANTIBODIES BE TREATED?

Management of the antiphospholipid syndrome is very controversial. Its therapy is not without side effects and only one-third of the patients with antiphospholipid antibodies ever experience a clinical problem. Over the years, my practice has evolved guidelines that I have found useful.

My patients with a positive test for the lupus anticoagulant or anticardiolipin antibody are told to take one baby aspirin a day. There is some preliminary evidence that this decreases the risk of thromboses. If a patient cannot tolerate even low-dose aspirin, all three antimalarial drugs used to treat SLE can prevent clots, so I prescribe one of them (Chapter 27). Aspirin can be combined with antimalarial therapy. However, patients who have had a thromboembolic event while taking aspirin or antimalarials require lifelong anticoagulation with warfarin (Coumadin).

The treatment of certain thromboembolic events requires hospitalization in order to use intravenous heparinization (or occasionally streptokinase), which dissolves the clots, followed by oral warfarin. The antiphospholipid syndrome demands a higher dose of warfarin than that used in other diseases since we're trying to achieve an international normalization ratio (INR), or blood-thinning level, of 2.5 to 3.5 as opposed to the usual 2.0 to 2.5. If the clots are arterial, the addition of a platelet antagonist such as dipyridamole (Persantine) may be helpful. Even though corticosteroids decrease levels of antiphospholipid antibodies or eliminate them from the blood, they promote clotting and do not necessarily decrease the risk of thromboemboli.

A few patients have what has been termed the *catastrophic primary antiphospholipid syndrome*. In other words, they experience repeated thromboembolic insults despite platelet antagonist therapy with aspirin, antimalarials or other agents, or therapeutic warfarin. This rare subset of patients can be difficult to treat, but giving chronic heparin intravenously or subcutaneously, with or without immunosuppression, is helpful.

## Summing Up

Approximately one-third of lupus patients have an antibody to phospholipids which, in the presence of certain predisposing factors unique to lupus, causes abnormal clotting. Individuals at risk are identified by performing antibody tests of phospholipids (especially to anticardiolipin antibodies), by using studies that check for prolonged clotting times, or by testing for syphilis and obtaining a false-positive result. Between 10 and 15 percent of those with SLE have clinical evidence of abnormal clotting, which can result in strokes, recurrent miscarriages, pulmonary emboli, and low red cell or platelet counts. There are few or no symptoms or warning signs. Few dietary or activity restrictions can aid prevention. Clotting complications can be prevented or minimized with platelet antagonist therapy (e.g., aspirin, antimalarials). Patients with thromboembolic events on platelet antagonists should be given warfarin, and a small subset of patients need chronic heparin.

# 22
# *Lupus Through the Ages: Lupus in Children and the Elderly*

Lupus comes in many sizes, shapes, and varieties unique to the age of onset, the presence of a specific autoantibody, and the appearance of certain clinical features. Most SLE—some 85 percent—occurs in individuals between the ages of 20 and 60. Does the disease manifest itself differently in youth and differently in those of more advanced age? The answer is yes. Four distinct types of SLE have been described that represent the remaining 15 percent of patients with the disease: neonatal lupus, lupus in childhood, lupus in adolescence, and older-age-onset SLE.

## NEONATAL LUPUS

At first, the concept that one can be born with lupus seems frightening. When a report appeared in 1954 that an infant was diagnosed with discoid lupus and its mother developed lupus shortly thereafter, the rheumatology community was concerned about the presence of such a chronic disease in infancy. However, between 1954 and 1992, only 300 such cases were described in the world's literature. Neonatal lupus is certainly quite rare.

If the neonatal period is defined as within 30 days of birth, how can one develop lupus during this period? In the 1950s, LE cells were shown to cross the placenta and produce positive LE cell results in healthy infants of lupus mothers for several weeks. It turns out the IgG but not IgM or IgA autoantibodies cross the placenta. Three IgG-containing autoantibodies that cross the placenta can damage fetal tissue: anti-Ro (SSA), anti-La (SSB), and anti-RNP (Chapter 11). Fortunately, all of these autoantibodies disappear by the eighth month after birth, since the baby is not able to make them. A review of the literature suggests that the prevalence of these autoantibodies in children with neonatal lupus is anti-Ro, 90 percent; anti-La, 53 percent, and anti-RNP, 2 percent.

### How Does a Physician Detect Neonatal Lupus?

The autoantibodies responsible for neonatal lupus settle in different tissues, especially the skin and heart. A review of published cases suggests that 54 percent of infants with neonatal lupus have congenital heart block, 37 percent have cutaneous lupus lesions, and 7 percent have both. Additionally, less than 8 percent have liver, gastrointestinal, blood, neurologic, or lung manifestations of lupus. Two-thirds of all reported patients are female.

*Congenital heart block* is the most serious complication of neonatal lupus. Anti-Ro and anti-La are attracted to fetal heart pacing tissue and can interfere with its development. Congenital heart block is defined by a slow fetal heart rate, evidence of heart block on an electrocardiogram (ECG) at birth, or abnormalities on a fetal heart echocardiogram (ultrasound). The death rate is 20 percent, but many babies can be saved by implanting a pacemaker. The incidence of congenital heart block in the general population is 1 in 20,000 births. Overall, the risk for mothers with SLE of having a child with congenital heart block is estimated at 1 to 2 percent. If they carry anti-SSA, anti-SSB or anti-RNP antibodies, the risk is about 5 percent; if they do not, it is zero.

*Cutaneous neonatal lupus* is manifested by the development of discoid or subacute cutaneous-type lesions during the neonatal period. The rashes, which disappear spontaneously after several months, respond to sun avoidance and steroid creams. Approximately 1 case of cutaneous neonatal lupus is found for every 30 mothers with SLE who carry the anti-Ro or anti-La antibody.

## What Kind of Disease Is Found in the Mothers?

Surprisingly, many women who have children with neonatal lupus feel perfectly healthy and do not have lupus. Of the 281 cases in the world's literature where maternal health has been ascertained, 40 percent had SLE, 38 percent had no disease, 13 percent carried a diagnosis of Sjögren's syndrome, and 9 percent had other autoimmune diagnoses. It should be remembered that 70 percent of all Sjögren's patients have anti-Ro. Why do healthy women have babies with neonatal lupus? Simply because one healthy women per thousand carries the anti-Ro or anti-La antibody. Many women are family members of patients with autoimmune diseases. Moreover, some healthy women develop clinically evident lupus or Sjögren's within several years of giving birth to a child with neonatal lupus, but most remain in good health.

## What Is the Outcome of Children with Neonatal Lupus?

With or without treatment, cutaneous lupus disappears within a few months. As mentioned above, the mortality rate of congenital heart block is 20 percent. Those who survive generally do quite well. Unfortunately, occasional reports have shown that some of the children born with neonatal manifestations of lupus develop SLE 10 to 15 years later. It is therefore strongly advised that all children with neonatal lupus be screened for SLE during their adolescent years.

### What Should Mothers with SLE Know about Neonatal Lupus?

Mothers with SLE who lack the anti-Ro, anti-La, or anti-RNP antibody are not at risk for delivering children with neonatal lupus. The 30 to 40 percent with lupus who carry these autoantibodies should be told that the risk of congenital heart block in their child is about 5 percent and of neonatal cutaneous lupus, 2 percent. Since cutaneous lupus is benign, the only precaution I take is to perform a fetal echocardiogram sometime between the 16th and 24th week of gestation, making sure that the fetal heart rate is not too slow. Fetal pacemakers can be implanted into the womb if necessary. I have never recommended that women with these potentially risky autoantibodies terminate a pregnancy, since a neonatal lupus risk of 1 in 14 is small and these children usually have normal lives.

## LUPUS IN CHILDREN

Somewhere between 5,000 and 10,000 children in the United States have lupus. Even though it looks like adult lupus under the microscope and is treated with the same medications, there are important conceptual and treatment considerations that enter into the overall equation.

First of all, lupus proportionately occurs more often in boys than men. Adult males make up 5 to 15 percent of the lupus population, but boys represent 20 to 40 percent of those with childhood lupus. Lupus is more severe at early ages and milder in older age groups. Whereas half of those with adult SLE develop organ-threatening disease (e.g., heart, lung, kidney, or liver involvement), 80 percent of those with childhood-onset SLE develop organ-threatening conditions. Nearly 70 percent of children with lupus have kidney disease, as opposed to 30 to 40 percent of adults. As a result, children with SLE require more aggressive monitoring and management. This becomes especially apparent since they are less likely to complain or to understand the seriousness of the inflammatory process. Fortunately, lupus is usually fairly easy to diagnose in children. The average lupus patient in the third decade has symptoms for 1 to 2 years before being diagnosed; in children, a diagnosis is usually made within 3 months.

The outcome of adult diseases is often discussed in medical journals and textbooks in terms of 5- or 10-year survival; in children, we are trying to achieve a 50-year survival. Since many of those with childhood SLE live out a normal life course, anything a doctor does has long-term implications. For example, corticosteroids may stunt growth and influence a child's stature for a lifetime. Chemotherapies may render a patient sterile and rarely cause cancer many years later which naturally has an impact on dating, career, and life-style choices. Certain medications used to treat lupus—such as the nonsteroidal anti-inflammatory indomethacin—are specifically ill-advised in young children. Other medications are not available as liquids or in strengths that can easily and safely be used by children. There are approximately 100 pediatric rheumatologists in the United States. All parents of children with SLE should have their pediatrician or healthcare-provider network seek out one of these practitioners for counsel and advice.

## LUPUS IN ADOLESCENTS

Pubertal teenagers with SLE have the same clinical manifestations as young adults with the disease. However, several psychosocial considerations are unique to adolescents.

Adolescents are notorious for not complying with prescriptions and medication. They find out very quickly what steroids do to their appearance and mood. Many would rather not take a medication that promotes fluid retention, acne, facial hair, easy bruising, and a puffy face. A fair percentage pretend to take their steroids but don't. Fragile social relationships with friends can be altered by sun avoidance, a skin rash, hair loss, swollen joints, and fatigue. They often have special concerns when confronted with issues of dating, marriage, or childbearing. All too often, physicians treat adolescents as adults and fail to address these issues. This mistake can be deadly. The importance of compliance with medication, regular blood monitoring, and being honest with the physician must be specifically spelled out and frequently reinforced. See Chapter 25 for a more detailed discussion.

## LUPUS AMONG THE ELDERLY

Two groups of individuals with SLE are found among patients over the age of 60: those who have had lupus for years and those who develop it for the first time. For those with a history of long-term SLE, the golden years are usually just that. Lupus tends to burn itself out after many years and rarely progresses after menopause. Patients still need to be careful in the sun and may have joint inflammation, but newly evolved organ-threatening disease is extremely rare.

The onset of lupus past the age of 60 is often very subtle. Our work has suggested that it takes an average of 3 years of symptoms before non-drug-induced SLE is diagnosed. Drug-induced lupus makes up most of the new lupus cases in older age groups, so a careful drug history should be taken. This topic is discussed in detail in Chapter 9. Several other diseases mimic SLE and differentiation can be difficult. In individuals over the age of 60, rheumatoid arthritis, polymyalgia rheumatica, and primary Sjögren's syndrome can all produce positive ANA tests, inflammatory arthritis, stiffness, aching, and elevated sedimentation rates. A rheumatology consultation is often helpful in arriving at the correct diagnosis. Late-onset SLE includes more males among its victims and is clinically manifested by aching, stiffness, dry eyes, dry mouth, and pleuritic pain.

Organ-threatening disease is quite rare and found in less than 20 percent of patients. Rashes, fevers, swollen glands, Raynaud's phenomenon, and neuropsychiatric lupus are also much less common than in young adults. Low-dose steroids, methotrexate, and antimalarials along with NSAIDs are frequently employed as treatment.

Less than 2 percent of all SLE occurs after the age of 70. Lupus in the elderly has an excellent outcome and deaths from the disease are extremely unusual.

## Summing Up

Lupus in newborns is very rare. Cutaneous neonatal lupus disappears spontaneously. Sometimes, the presence of congenital heart block may require permanent treatment with a pacemaker or medication that controls the heart's rhythm, but this condition usually has a favorable outcome. Recent reports suggest that a minority of neonatal lupus patients may develop lupus during adolescence. Childhood lupus is frequently serious and organ-threatening. Aggressive management increases the chances for a good quality of life during adulthood. The teen years are fraught with compliance problems and psychosocial complications that require special attention. For those who develop symptoms of SLE after age 60, drug-induced lupus and other forms of inflammatory arthritis must be ruled out first as possible culprits. Senior citizens with lupus usually have mild, nonprogressive disease.

# 23
# *Is It Really Lupus?*

## DIFFERENTIAL DIAGNOSIS AND DISEASE ASSOCIATIONS

Ten million Americans have a positive lupus blood test, but probably less than one million have lupus. Many people fulfill one, two, or three American College of Rheumatology (ACR) criteria for SLE but lack the requisite four criteria reviewed in Chapter 2. What do they have? Are other autoimmune diseases present? And while we're on the subject, are there any other disorders that lupus patients tend to get or are spared from? When your physician considers the presence of other similar diseases, this decision process is called *differential diagnosis*. This chapter will put these issues into a practical, logical perspective.

## HOW CAN WE TELL IF IT'S LUPUS?

If a person's rheumatic complaints do not fulfill the criteria for systemic lupus or any other rheumatic disease, what is the problem and how can the doctor tell for sure? First of all, cutaneous (discoid) lupus and drug-induced lupus have definitions separate from those for SLE, which are reviewed in Chapter 2. A host of patients with various forms of lupus do not always fulfill the ACR criteria even when the diagnosis is self-evident. For example, evidence from biopsies of the kidney will provide clear-cut proof of lupus if the kidney is affected by the disease.

The average person with symptoms of lupus can take 1 or 2 years to be diagnosed because the full-blown disease is not present. We call this evolution by its Latin term, *forme fruste lupus*. Similarly, a small percentage of patients with *palindromic rheumatism* develop SLE over a 5- to 10-year period. A palindrome is a word or phrase that is the same spelled forward or backward, such as "Madam, I'm Adam" or "Dad." Patients with palindromic rheumatism can be fine one day and have rashes or swollen knees the next. Three days later they are fine, with no hint of there ever having been a problem. These flareups are cyclical and blood tests are often negative. Half of these patients go on to develop rheumatoid arthritis, but some become patients with lupus.

What if lupus-like symptoms have appeared for years? Many patients have nonspecific complaints of fatigue or aching along with a positive ANA. Most immunology laboratories have ANA panels that can be performed. As discussed in Chapter 11, certain autoantibodies are simply not seen in healthy people. If any patient seeks my consultation and has any of the following abnormal tests with the above complaints, I know that a real immunologic disorder exists: anti-DNA, anti-Sm, anti-RNP, C3 complement, C4 complement, high Westergren sedimentation rate, positive rheumatoid factor, high CPK, false-positive syphilis test, anticardiolipin antibody, antineuronal antibody, antihistone antibody, antiribosomal P antibody, anti-Ro (SSA), anti-La (SSB), or a broad gamma globulin band on serum protein electrophoresis. *In questionable situations, the presence of these antibodies must be confirmed by a second, independent laboratory.* Some of the tests are difficult to perform and occasionally false-positive results are obtained.

What can a doctor do to confirm the disease in a suspected lupus candidate who has a positive ANA, lupus-like symptoms, and negative tests for other antibodies? If joint or muscle complaints are prominent, lupus can be distinguished from fibromyalgia (discussed below) by a *bone scan.* Lupus can inflame the joints; fibromyalgia does not. In lupus, a bone scan may pick up increased blood flow to the joints as well as bone and muscle inflammation, whereas fibromyalgia will produce a normal bone scan. Inflammatory lupus arthritis responds to a group of medications different from those used in noninflammatory fibrositic arthralgias. Another useful procedure is the *lupus band test.* Its applications are reviewed in detail in Chapter 12, but in essence the presence of a specific combination of immune reactants at the junction of the dermis and epidermis under the skin is seen only in SLE even if there is no rash. Finally, half of all first-degree relatives (parents, siblings, or children) of SLE patients demonstrate a positive ANA on blood testing even though only 25 percent will ever develop an autoimmune disease and less than 10 percent will become lupus patients. Nonspecific symptoms in these individuals should be followed closely.

The biggest mistake a doctor can make is to classify a patient as having SLE when he or she doesn't have the disease. Occasionally, I come across what I can only call a "lupus wanabee." These individuals are convinced they have the disease on the basis of reading and talking to their friends. It is very difficult to convince them that they don't. Some want to be diagnosed as having lupus perhaps for psychological reasons: they want loved ones to pay more attention to them or feel sorry for them, or they want to prove to family members that they are not "crazy." Some may eventually find a doctor who will agree with their self-diagnosis and treat them. But this is not the sort of illness that can be placated with innocuous and harmless treatments. The therapy for SLE involves toxic, expensive, and time-consuming treatment. Disease-modifying therapies (anything other than an NSAID) should never be given unless a firm diagnosis has been made or an organ-threatening complication is clinically evident. Labeling a patient with SLE who does not have the disease can make it difficult for him or her to obtain gainful employment, health insurance, or life insurance and can mean lifelong stigmatization.

## WHAT IS ANA-NEGATIVE LUPUS?

Heidi was sure she had lupus even though six different doctors had obtained six different negative ANAs in six different labs. She had a butterfly rash on her cheeks, was tired and achy, and had a checkerboard mottling (called livedo reticularis) on her legs. Finally, the third rheumatologist she saw was sympathetic to her predicament. Even though her blood chemistry profiles, chest x-ray, electrocardiogram, sedimentation rate, CPK, anti-DNA, and complement levels were negative or normal, Dr. Schwartz obtained additional blood for testing. Heidi had a positive anticardiolipin antibody and a false-positive syphilis serology. These antibodies are seen in 20 percent of discoid lupus patients and are associated with livedo reticularis. Dr. Schwartz also referred Heidi to a dermatologist, who did a lupus band test on her cheek rash; it came back positive. A bone scan showed increased blood flow to her hands and feet, suggesting an inflammatory arthritis. Dr. Schwartz diagnosed Heidi as having ANA-negative lupus since she fulfilled the criteria by having arthritis, sun sensitivity, discoid rashes, and a false-positive syphilis serology. Three years later, Heidi's ANA blood test became positive.

Until 1985, 10 percent of all lupus patients had a negative ANA test. The introduction of improved testing material for performing the ANA test has decreased this percentage to 3 percent. Between 1980 and 1989, my office treated 464 patients who fulfilled the ACR criteria for lupus; seventeen of them were ANA-negative. In analyzing this group, we found that patients fell into four basic categories. One-third had antiphospholipid antibodies and one-third had biopsy-documented kidney lupus. Of the remaining third, half ultimately became ANA-positive. The last group had advanced disease; prolonged treatment with steroids and chemotherapy made their ANA disappear. A variety of rarer causes of ANA-negative lupus exist, such as the presence of anti-Ro (SSA) antibody without ANA. Nevertheless, if a patient does not fall into any of these four categories, some of the lupus-related or lupus-mimicking disorders discussed in this chapter might be the culprit.

## WHAT NONAUTOIMMUNE DISORDERS MIMIC LUPUS?

Many diseases mimic SLE; they fall into several broad general categories. For example, almost every disorder of *hormonal imbalance*—from thyroid abnormalities to evolving menopause to diabetes—can appear with symptoms such as fatigue, aching, and feeling feverish that may lead the doctor to order lupus blood tests. *Blood* or *tissue malignancies* ranging from lymphoma to breast cancer can give positive ANAs and induce constitutional complaints. *Infectious processes,* especially from viruses, are associated with positive ANAs as well as lupus-like symptoms. All these possibilities should be ruled out before a diagnosis of lupus is made.

## HOW CAN WE DIFFERENTIATE LUPUS FROM OTHER RHEUMATIC DISEASES?

Especially during the first year of symptoms, rheumatoid arthritis, scleroderma, mixed connective tissue disease, inflammatory myositis, and other forms of systemic vasculitis can be very difficult to differentiate from each other. It is not usually critical to make the diagnosis at first, since steroids and immune-suppressive therapy can be used to treat critical complications of any of these disorders.

*Rheumatoid arthritis* (RA), which afflicts 4 million Americans, is the only autoimmune rheumatic disease that is more common than lupus. In its initial presentation, RA can be difficult to tell from lupus; but within several months, the diagnosis usually becomes obvious. One-quarter, or 25 percent, of patients with lupus have a positive rheumatoid factor and 25 percent with rheumatoid arthritis have a positive ANA. The hallmark of RA is an autoimmune reaction in the synovium, the tissue lining the joint. It is uncommonly complicated by systemic organ involvement, and deforming rheumatoid joint disease causes erosions (actual bone destruction) that resembles a mouse bite on x-rays. Lupus induces joint deformities less than 10 percent of the time and erosions are almost never seen. An occasional patient has distinct features that fulfill the ACR criteria for both RA and SLE. Called *rhupus,* this rare disorder is managed with drugs useful for both diseases, such as nonsteroidals, steroids, methotrexate, and antimalarials.

Most patients with *scleroderma,* an autoimmune disease characterized by inflammation that heals with tightening of the skin and scarring of the tissues, also have a positive ANA. The 500,000 to 750,000 Americans with scleroderma-related disorders (e.g., progressive systemic sclerosis, CREST syndrome—or *Calcinosis, Raynaud's Esophagitis, Sclerodactyly, Telangiectasia*—autoimmune Raynaud's, mixed connective tissue disease) are frequently misdiagnosed as having SLE. A few lupus patients fulfill criteria for both diseases (called *lupoderma*), and in most cases the condition eventually evolves into a pure scleroderma. In reality, a process known as *mixed connective tissue disease* (*MCTD*) probably makes up the majority of these cases. By definition, as well as being ANA-positive, MCTD patients must have a positive anti-RNP. Their condition resembles lupus, but they tend to have puffy hands, complain of heartburn and swallowing problems, and have interstitial scarring of the lungs on chest x-ray. Raynaud's phenomenon is almost universally seen in MCTD. Our survey of 464 patients with SLE shows that 25, or 5 percent, also met the definition for

MCTD. It used to be taught that MCTD was a more benign process that spared the kidney, central nervous system, and blood, but this is not true in children and adolescents. MCTD is managed in the same way as SLE.

*Inflammatory myositis,* as in dermatomyositis or polymyositis, is a feature of lupus in 10 to 15 percent of patients. However, the very high CPKs seen in dermato- or polymyositis, its distinct skin papules, and its association with malignancy and heliotrope-like rashes are not observed in SLE. In other words, lupus is occasionally characterized by a mild, bland, frequently asymptomatic muscle inflammatory process.

Rheumatoid arthritis, lupus, scleroderma, inflammatory myositis, and MCTD all involve inflammation of the small and medium-sized arteries and arterioles. *Polyarteritis nodosa* is a primary vasculitis of those caliber vessels. It can mimic lupus and most patients have a positive ANA. The diagnosis of polyarteritis is confirmed by a high-titer positive p-ANCA with a special staining, a blood test fairly specific for systemic vasculitis, along with evidence of inflammation of the vessels documented by an angiogram or biopsy. A group of patients have features of two to six of the disorders mentioned above, but do not seem to fit perfectly into any one of them. They are said to have a *crossover syndrome,* or an *undifferentiated connective tissue disease* process. Many individuals initially exhibit a crossover syndrome that evolves into a distinct, definable disorder over time. Occasionally, rare forms of vasculitis affect medium and small-sized vessels, which heal with scars known as granulomas; these are difficult to differentiate from lupus at first. *Wegener's granulomatosis* and *Churg-Strauss granulomatous vasculitis* are ultimately diagnosed by the pathologic finding of granulomas or a positive c-ANCA blood test in Wegener's. Finally, an unusual form of vasculitis called *Behcet's* is rarely diagnosed as ANA-negative lupus. Its features of mouth ulcers, eye inflammation, and central nervous system involvement resemble SLE, but the lack of a positive ANA usually leads physicians to a correct diagnosis. Finally, as mentioned in the last chapter, older people can develop an aching in the hips and shoulders with severe stiffness and a high sedimentation rate. Some of these individuals, who are ultimately diagnosed with *polymyalgia rheumatia,* have positive ANAs as a function of age, and lupus must be ruled out.

Several features of autoimmune disease can also appear by themselves without fulfilling criteria for any of the disorders above mentioned. These include Raynaud's phenomenon, idiopathic thrombocytopenic purpura (ITP), and Sjögren's syndrome in addition to those listed in the above paragraph.

## DOES LUPUS DECREASE THE RISK OF GETTING OTHER DISEASES?

For reasons that are not clear, several disorders have a ''negative'' association with lupus—in other words, having lupus decreases your chances of getting certain diseases. These include *amyloidosis, sarcoidosis, ankylosing spondylitis,* and *AIDS*.

Since 500,000 to 1 million Americans had SLE and at least 50,000 of them received blood transfusions between 1978 and 1983 when our blood supply was not safe, one would have expected some of the recipients to develop *AIDS,* a disease caused by the *human immunodeficiency virus*. In fact, not a single case was reported. (The 50,000 figure is derived from adding the number of lupus patients on dialysis who would have been given frequent transfusions, the number of orthopedic surgeries such as hip replacements for avascular necrosis, lupus patients with autoimmune hemolytic anemia, and the common use of transfusions over a 5-year period in that era for gynecologic surgery. With the advent of the drug EPO, we rarely give transfusions to patients on dialysis, and patients can now donate their own blood prior to surgery.)

Even though 90 percent of AIDS patients in the United States are males and 90 percent with SLE are female, statistically one would still expect several hundred cases of AIDS with SLE. However, at the time of this writing, only six cases have been reported in the world's literature. Four were children with congenital AIDS who developed SLE, and one lupus patient received a kidney transplant from a human immunodeficiency virus (HIV)-positive donor.

Does lupus protect patients from AIDS? Since they represent opposite poles of the immune system (AIDS patients have few CD4 cells and lupus patients have overactive CD4 cells), this possibility is an intriguing one which deserves further study. Interestingly, I have occasionally come across lupus patients with false-positive HIV tests; this phenomenon may reflect the ability of SLE to produce antibodies to many different viruses. Nevertheless, lupus patients must be cautioned not to let their guard down, to practice safe sex, and to avoid street drugs.

## WHAT IS THE RELATIONSHIP BETWEEN LUPUS AND FIBROMYALGIA?

Fibromyalgia, or fibrositis, has repeatedly been mentioned throughout this book. Here's where you find out just what it is.

### What Is Fibromyalgia?

Fibromyalgia is a disorder that afflicts 6 million Americans. About 90 percent of them are women, and most develop the disease between the ages of 20 and 50. It is a pain-amplification syndrome characterized by widespread stiffness and aching of at least 3 months duration. In order to fulfill the ACR criteria for fibromyalgia, one must have tender points (defined as wincing or withdrawing in pain when 4 kilograms or 9 pounds of pressure is applied) in at least 11 of 18 designated points in all four quadrants of the body (right side, left side, above the waist, and below the waist, as shown in Figure 13). If tender points are found in

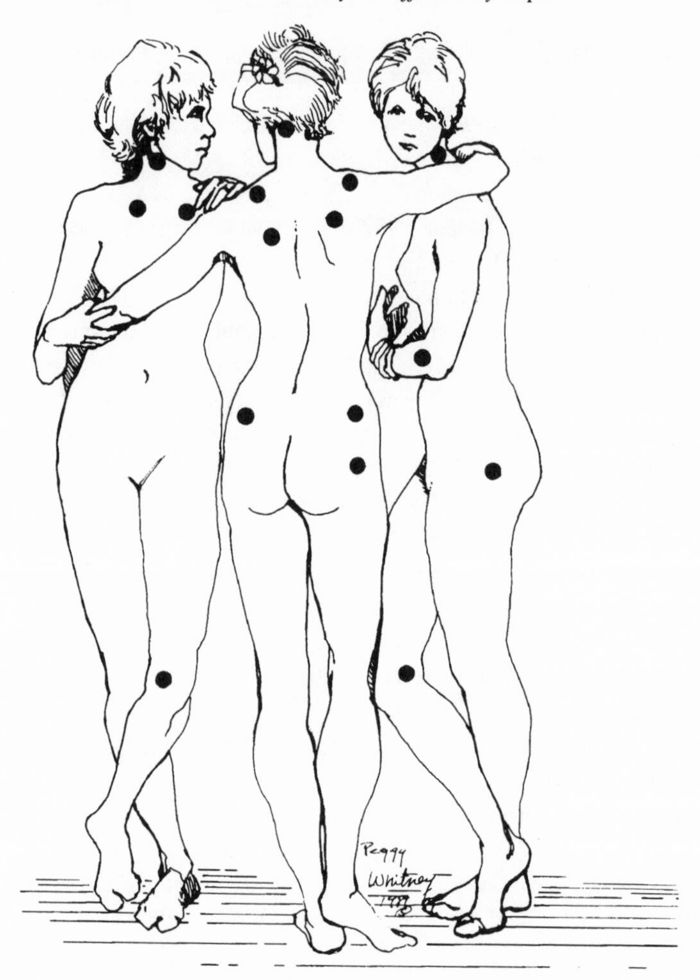


**Fig. 13.** *Tender Point Locations in Fibromyalgia*

fewer than four quadrants, the patient does not have fibromyalgia but a *regional myofascial syndrome*. Most of the tender points are in the upper back and neck area, buttocks, and chest. In addition to musculoskeletal symptoms, patients with fibrositis may complain of nonrestorative sleep (waking up in the morning after lying down for 8 hours but not feeling refreshed); functional bowel symptoms such as abdominal cramping, bloating, and distension; a sensation of swelling with numbness and tingling, profound fatigue, and occasionally cognitive dysfunction. These symptoms are similar to those discussed in Chapter 15.

The cause of fibromyalgia is unknown, but pain amplification probably results from a dysfunction of neurotransmitters (such as epinephrine, dopamine, or serotonin) and cytokines (Chapter 5) and their relationship to pain chemicals including endorphins, enkephalins, and substance P. Fibromyalgia can be brought on by trauma, a virus, or an inflammatory disorder, among other causes, or it may "just happen" as a reaction to physical or emotional stress.

## How Many Lupus Patients Have Fibromyalgia?

Several surveys have suggested that about 20 percent of all lupus patients also fulfill the ACR criteria for fibromyalgia. The most common causes of fibromyalgia in SLE are a reaction to active musculoskeletal lupus, temporary fibromyalgia after a viral infection or trauma, poor coping mechanisms, and a steroid withdrawal syndrome. The latter occurs, for example, when your doctor sees how well things are going and decreases prednisone doses from 20 to 15 milligrams a day. Decreased steroid doses can result in a flareup of muscle and joint aching without a worsening in laboratory testing, objective synovitis, or swelling indicating a lupus flare. These symptoms often lead patients to call their doctors, who then raise the prednisone back to 20 milligrams daily. This may be a mistake. These complaints represent withdrawal symptoms and will disappear spontaneously over a 1- to 3-week period. Not only do alterations in steroid doses aggravate fibromyalgia, but the skin of all patients on corticosteroids develops an increased sensitivity to pressure which imitates fibrositis.

## How Can We Tell Lupus from Fibrositis?

Differentiating active lupus from fibromyalgia is critical. This is often made difficult by two confounding factors: Lupus patients can have concurrent fibromyalgia, and 10 percent of fibromyalgia patients have a positive ANA. In patients with SLE, complaints of fatigue, muscle aching, and stiffness can represent active lupus or fibrositis. If a recent infection, trauma, or steroid tapering is ruled out, active lupus is usually detected by the presence of a rash or swollen joints on physical examination or is evident in tests for anemia, elevated sedimentation rate, a high anti-DNA, or low complement levels. A common mistake among internists and even rheumatologists is to take a patient's symptoms at face value and treat them with toxic medication even though no laboratory parameter validates that a lupus flareup is present. Fibrositis flareups respond only temporarily to rises in corticosteroids, and the patient may actually feel worse within several weeks.

## How Do Doctors Treat Fibromyalgia?

Fibromyalgia is managed in several ways. First, doctors reassure patients that even though it's a real disease, it is neither life-threatening nor crippling. Rheumatologists usually provide articles from the Arthritis Foundation or the American Fibromyalgia Syndrome Association that document its nonprogressive nature. Second, physicians encourage patients with fibrositis to adjust their lifestyles in a way that will ensure restful sleep, pacing of time, and improved coping mechanisms. Counseling may be recommended. Also, physical measures such as moist heat, gentle massage, biofeedback, and coolant sprays followed by muscle stretching (called spray and stretch) are also employed. An occasional injection of a local anesthetic (with or without a local steroid) can be used at the trigger point of pain. A patient's work station and job description are analyzed in an effort to minimize alterations in body mechanics that could irritate muscles and stress joints. Electronic acupuncture (e.g., acuscope, neuroprobe, TENS units) may also be helpful.

Finally, numerous medications can be tried. They include NSAIDs, which are modestly helpful. Muscle relaxants such as Norgesic, Parafon, and Soma are beneficial for a few hours. Tricyclic antidepressants—such as cyclobenzaprine (Flexeril), doxepin (Sinequan), amitriptyline (Elavil), and nortriptyline (Pamelor)—promote restful sleep, relax muscles, minimize reactive depression, and raise pain thresholds. Benzodiazepines such as diazepam (Valium), clonazepam (Klonopin), and chlordiazepoxide (Librium) do the same thing but are potentially addictive and are usually reserved for the 15 percent with fibrositis who cannot tolerate tricyclics. Serotonin agonists, including fluoxetine (Prozac), sertraline (Zoloft), veulafaxine (Effexor), and paroxetine (Paxil), can be helpful alone or along with tricyclics. They may worsen sleep habits, but with a tricyclic they raise pain thresholds as well as decreasing fatigue when taken in the morning. Fibromyalgia is a chronic process whose course waxes and wanes, but most patients respond to treatment.

## Summing Up

Lupus is often difficult to diagnose. If the ACR criteria are not fulfilled, additional blood tests, bone scanning, or a lupus skin biopsy band test can assist in making the diagnosis. ANA-negative lupus is rare, and a physician confronted with this diagnosis should embark on a workup to exclude other diseases that mimic lupus, such as scleroderma, rheumatoid arthritis, vasculitis, polymyalgia rheumatica, fibrositis, or Behcet's syndrome. Low-titer positive ANAs are found in patients with malignancies and infections, which must be ruled out before the diagnosis of lupus becomes established. Most autoimmune disorders have overlapping features with SLE and must be considered, since their management may substantially differ. Finally, lupus patients are at an increased risk for having a concurrent fibromyalgia, which is also managed differently from SLE. Sorting out lupus from fibrositis flareups presents a major challenge for patients and their healers.

# Part V
# THE MANAGEMENT OF LUPUS ERYTHEMATOSUS

Having read about the various symptoms, signs, and laboratory features of lupus, we must now consider ways to treat the disorder. I've taken a therapeutic approach that will help patients feel better by working with their physicians and health-care team, by showing them how to maximize coping skills, and by promoting an understanding of the rationale behind specific treatment plans.

The treatment of lupus erythematosus is divided into four categories: physical measures, medication, surgery, and counseling. All four are closely interrelated, although surgery plays a minor role in the management of SLE. Simply stated, "The head bone is connected to the lupus bone." A doctor might prescribe all the correct medications, but if emotional stress overcomes the patient's will to recover, it could all be for naught. We review these areas in this section. Feel free to skip around or look up in the index any specific treatment feature that might interest you, but remember: the treatment of lupus is multifaceted and will be unsuccessful unless all four categories are given careful attention.

# 24
# *How to Treat Lupus with Physical Measures*

Let's look at physical measures first. We have a fair amount of control over these management techniques, and to some extent, controlling the environment represents a commonsense approach. The physical or environmental factors we are going to discuss include the effect of sunlight, diet, exercise, heat, rest in the treatment of fatigue, and the impact of weather.

## DO LUPUS PATIENTS REALLY NEED TO AVOID THE SUN?

The sun emits ultraviolet radiation in three bands known as A, B, and C. Only the first two, ultraviolet A (UVA or "tanning") and ultraviolet B (UVB or "burning"), are harmful to lupus patients. (The mechanisms by which sun damages the skin and aggravates lupus are discussed in Chapters 8 and 12.) Many of my lupus patients say the sun does not bother them and ask if they really need to avoid it. On the other hand, another group of my patients are so sun-sensitive that they develop a rash along with fatigue and aching even when they are exposed to open, uncovered fluorescent lights.

The truth about sun exposure lies somewhere in between. When rheumatologists sent lupus patients questionnaires about how they feel in the sun, 60 to 70 percent replied that they avoid the sun because it gives them a rash or makes them feel tired, achy, or feverish. However, when dermatologists administered ultraviolet light to a small, defined area of skin and later biopsied it to look for inflammation or irritation, they found that only 30 percent of their patients with SLE had reproducible light sensitivity. The reason for this discrepancy is that ultraviolet light damages the skin in a time- and dose-related fashion. I have patients who tell me that they can tolerate 15 minutes of sun exposure but begin to feel sick after 20 minutes. Ultraviolet light is present *even* on a cloudy day: UVA is constant throughout the day, but UVB (which is more harmful in lupus) is strongest between the hours of 10 A.M. and 3 P.M. (standard time). Ultraviolet light is more powerful at higher altitudes, and it can also be reflected on certain surfaces, such as sand and snow. I advise my sun-sensitive patients to undertake their necessary outdoor activities in the early morning or late afternoon, so they can avoid the peak UVB period. Medications that increase one's sensitivity to the sun include most sulfa-containing antibiotics and certain tetracyclines.

Do sunscreens help? In most cases they can be useful, but an understanding of how they work is important. Most of the commercially available sunscreens are rated on a scale known as SPF (sun protection factor). An SPF of 15, for example, means that you are 15 times more protected than with no protection. SPFs below 15 are of little value in lupus, and those over 30 may cause the skin to dry, burn, sting, or itch. These ratings apply only to UVB light; some commercially available preparations also block UVA. Two products with both UVA and UVB blocks include Shade UVA Guard and Ti-Screen, which are available in 15- and 30-SPF strengths. Sunscreens are "over-the-counter" preparations, which means that a prescription is not necessary. If you are going to be out in the sun for 5 minutes or less, protection usually is unnecessary. For longer periods of sun exposure, a sunscreen can be applied every 2 to 3 hours to any uncovered area, especially the face. Protective clothing and wide-brimmed hats are also useful.

A small subset of my patients (less than 5 percent) are extremely sensitive to ultraviolet light. Most of them carry the anti-SSA (Ro) antibody (Chapter 11). Sun-sensitizing chemicals are found in certain perfumes, mercury vapor lamps, xenon arc lamps, halogen or tungsten iodide light sources, and photocopy machines; excessive exposure should be avoided. Fluorescent lighting rarely presents a problem if the fixtures have a covering. Sleeves that block UV emanation without reducing illumination from fluorescent lighting are available (Solar Screen Company, Corona, NY). Tinting car windows and wearing special protective sunglasses may be advisable. Even lupus patients who are not sun-sensitive must be aware of the potential damage of UV light and should take precautions.

Despite these precautions, the lupus patient who exercises prudence and caution need not become an "environmental cripple." Lupus patients should approach the issue of ultraviolet light with common sense and not obsessiveness or panic.

## IS THERE A DIET FOR LUPUS?

Individuals with SLE should eat a well-balanced, healthy, nutritious diet. Current recommendations call for a diet consisting of 50 to 55 percent carbohydrates, 15 percent protein, and less than 30 percent fat. Diet books for arthritis are a multimillion-dollar industry, and one of the questions most commonly asked of rheumatologists deals with the role of diet in lupus. It might seem surprising, but few nutritional modifications apply to SLE. Things that do affect lupus can be divided into two categories: factors that are lupus-related and those that are medication-related.

For starters, fish oil has anti-inflammatory properties. This has been documented in patients with rheumatoid arthritis and in animal models of SLE. Eating several fish meals a week is equivalent to taking several extra aspirins. It will never cure the disease but might bring about a modest improvement in well-being. Fish oil capsules are appropriate substitutes, but they can irritate the stomach, and it takes 8 to 10 capsules a day to substitute for one fish meal.

One food supplement to stay clear of is alfalfa sprouts. They contain an amino acid known as L-canavanine, which increases inflammation in patients with autoimmune disease. All members of the legume family contain L-canavanine, but it is highly concentrated in alfalfa sprouts. Well-documented flareups of lupus disease have been associated with increased consumption of alfalfa sprouts and have disappeared when sprouts are avoided. Alfalfa is an ingredient in many food products, and some aggressively marketed ''natural'' vitamin remedies contain alfalfa. Such products (e.g., Km) should probably be avoided by patients with SLE. I advise my patients to bring me copies of the labels on health-food packaging to make sure the products don't contain any ingredients that might be harmful in SLE.

Numerous medications are used to treat SLE, but only one has any dietary implications. Corticosteroids can raise blood sugar, serum cholesterol, and triglyceride levels and increase blood pressure. Therefore, steroid-dependent patients who require a dose of more than 10 milligrams of prednisone a day should decrease their sugar, salt, and fat intake.

## WHAT ABOUT VITAMINS?

No specific vitamin is recommended for lupus, but under special circumstances certain vitamins may be useful. For example, vitamin $B_{12}$ and folic acid treat some of the anemias seen in SLE patients, vitamin $B_6$ has a mild diuretic effect, and vitamin D derivatives play a role in managing specific types of osteoporosis (thinning of the bones) that are observed in the disease (Chapter 13).

## CAN LUPUS PATIENTS EXERCISE?

Judicious exercise is a very important part of managing lupus. It can strengthen muscles, improve flexibility, and promote a sense of well-being. Inactivity can promote osteoporosis, muscle weakness, and wasting. Patients who are not fit are less able to respond to various stresses in the environment.

The optimal conditioning program involves engaging in activities that strengthen muscle tone and improve endurance without putting too much stress on a single joint. Isometric exercises that involve contracting muscles without moving the involved joint are good ways to start. Later on, walking, swimming, or bicycling are excellent activities. To start, one can take a 5-minute walk twice a day and build up to an hour long walk three to five times a week. Limits of endurance are highly variable among individuals. The "talk test" asks: Can you talk comfortably while exercising? If not, it's best to slow down or stop. Also, one should *never* exercise beyond the point of minimal discomfort.

When a joint or muscle is painful, local heat can be applied. Moist heat (e.g., shower, bath, hot tub, jacuzzi, thermophore) is superior to dry heat. If an area is acutely injured, the application of ice will minimize swelling during the first 36 hours. Inflamed joints must not be exercised. This can be harmful. For example, engaging in such activities as tennis, bowling, golf, weight lifting, or rowing with an inflamed or swollen hand, wrist, or shoulder can aggravate the disease process. But the injured area should not be ignored completely. An inflamed joint should be put through its full range of motion several times a day. This helps prevent contractures and muscle atrophy.

## WOULD A REHABILITATION PROGRAM HELP?

As discussed in Chapter 13, the inflammatory arthritis of lupus causes visible swelling of the joints in 20 to 30 percent of patients with the disease and deformities in less than 10 percent. Patients with inflammatory arthritis often benefit from a formal rehabilitation program.

*Physical therapists* are licensed allied health professionals (look for RPT, registered physical therapist, after the name). They help improve conditioning, instruct you on how to move inflamed joints without damaging them, and will introduce you to muscle-strengthening regimens. They can also apply hot packs, administer ultrasound, give gentle massages, and employ spray-and-stretch techniques (using a coolant spray followed by gentle tissue stretching) for tender fibrositic tissues. A RPT will be glad to suggest a conditioning program to your doctor.

*Occupational therapists* work with physical therapists. They are also licensed allied health professionals and provide valuable, underutilized expertise (look for OTR, registered occupational therapist, after the name). They will perform an "Activities of Daily Living" (ADL) evaluation. After examining what an individual does in the course of a day at work and at home, the therapist can suggest methods of energy conservation and joint protection. In other words, is there a way to cook a meal or get on or off a toilet seat with the least amount of stress on an inflamed or damaged joint, which also can minimize discomfort? An OTR is expert in recommending assistive devices such as splints or braces and practical modifications in the workplace or home (e.g., special spoons, toothbrushes, shoehorns) that make life easier. Consider this example of how an OTR works:

> Kim is a paralegal for a large law firm. Her employers provided her with a work station that enabled her to do word processing, answer telephones, and file records. But she soon began complaining of neck and upper back pain as

well as numbness and tingling in her hands. Although she has had SLE for 4 years, she never had any musculoskeletal problems other than occasional aching until she took this job. Her doctor ordered an occupational therapy consultation, which carefully evaluated her work station. The OTR recommended a higher chair with a firmer back and a swivel screen for the computer. She instructed Kim to type for no more than 20 minutes at a time. After implementing the changes, Kim began to feel better. The OTR showed Kim how to squat rather than stoop when picking up a file weighing more than 10 pounds, and how to lift the files to equalize the weight on both sides of her body. Her doctor suspected that she might have carpal tunnel syndrome, which would cause numbness and tingling in her hand. He injected the carpal tunnel with cortisone and she wore wrist splints provided by the OTR at night for several weeks. Kim is now pain-free.

*Vocational rehabilitation* counselors provide job training for patients who are unable to continue working at their current jobs. Their services are usually obtained through worker's compensation and disability insurance programs. Some examples of vocational rehabilitation candidates are sun-sensitive farmers or fishermen or office employees with hand deformities who are unable to type.

*Psychologists* as well as physical and occupational therapists assist lupus patients in learning relaxation techniques that promote improved sleep habits and reduce stress. Some of these techniques include biofeedback, yoga, gentle massage, and hypnosis. (Chapter 25 discusses stress and coping mechanisms, along with some of these techniques, in greater detail.)

## WHY ARE LUPUS PATIENTS ALWAYS TIRED?

One of the most common complaints I hear from my patients is that they are always tired. There are many reasons for this. Lupus is associated with anemia and active inflammation, both of which promote fatigue. Some of the medicines that doctors prescribe for high blood pressure and inflammation, for example, can make one drowsy. Alternatively, the stresses of dealing with a serious disease and its associated depression are also exhausting. Many of my lupus patients whose blood tests show no active disease deny that they are depressed but still complain of fatigue. A partial explanation is on the horizon. Some recent evidence shows that a group of glycoproteins known as *cytokines* are associated with fatigue (Chapter 5), and cytokine dysfunction is an established feature of lupus.

The best way for a lupus patient to manage fatigue is to follow this course of action:

1. Determine the cause of fatigue. Ask your doctor to undertake an evaluation for reversible causes of fatigue such as anemia, hypothyroidism (low thyroid levels), elevated blood sugars, or a lupus flare. If one of these is the cause, it can be treated.

2. Pace yourself. Keeping active prevents the cytokines from getting the better of you. Staying in bed all day only increases fatigue. Commit yourself to an hour or two of activities followed by a rest period of 15 to 20 minutes. Repeat this several times a day. Most lupus patients can perform 8 to 10 hours of productive work a day if they alternate periods of activity with periods of rest. Working 6 hours straight can be traumatic; it promotes a feeling of exhaustion that may call for several days of recovery.
3. Take appropriate medication. Many lupus medications decrease fatigue, in particular corticosteroids and antimalarials. Your physician may wish to prescribe certain other medications to treat fatigue. Depending on the circumstances, some of these include iron, thyroid, serotonin reuptake inhibiting antidepressants (Prozac, Zoloft, Paxil), and tricyclic antidepressants (Elavil, Sinequan, Pamelor).
4. Get a restful night's sleep. Many lupus patients have a secondary fibrositis which is associated with sleep disturbance. Not getting a restful night's sleep saps your energy and promotes fatigue. Tricyclic antidepressants relax the muscles and help induce restful sleep without being habit-forming or dangerous.
5. Walk. A conditioning program with aerobic exercise such as walking gets more oxygen into the tissues, strengthens muscles, and will give you a sense of well-being while also reducing fatigue.

## CAN WE BLAME IT ON THE WEATHER?

It has been jokingly suggested that patients with rheumatic disease make excellent meteorologists. Changes in barometric pressure frequently lead to symptoms of increased stiffness and aching in the joints. In other words, when the temperature goes from hot to cold or the humidity from dry to wet, lupus patients complain of a feeling of stiffness. But patients with lupus needn't worry about their weather predicting abilities. Fewer symptoms are observed in patients who live in desert climates, where the temperature and humidity are consistent. Those who live in the midwestern United States, for example, might be a little more achy or stiff on certain days. When traveling to different parts of the country, they may find that it takes a day or two to acclimate to their new environment.

### Summing Up

A good deal of lupus treatment involves things in the environment that can be controlled. A lot can be accomplished by avoiding the sun, eating a healthy, well-balanced diet, engaging in a moderate amount of general strengthening and conditioning exercise, pacing oneself, and avoiding frequent changes in barometric pressure. Now let's continue with things patients can do for themselves. The next chapter teaches you how to cope better.

# 25
# *You Can Help Conquer Lupus*

When confronted with a diagnosis of lupus, most patients are initially frightened about the prognosis. Often, their first reaction is to ask what they did wrong. Coping with the diagnosis of lupus can be a difficult proposition. So many aspects of the disease must be dealt with that can at times seem overwhelming. Studies have shown that over half of lupus patients express a broad range of feelings including stress, anger, depression, fear, guilt, and pain. Conquering these concerns is so complicated that the Arthritis Foundation (with the cooperation of The American Lupus Society and the Lupus Foundation of America) supports what is called SLESH, or the Systemic Lupus Erythematosus Self-Help course, which consists of seven weekly two-and-a-half-hour sessions addressing these issues.

This chapter offers an overview of the principal psychosocial problems that lupus patients encounter and formulates meaningful, constructive approaches for dealing with them. Active lupus and medications given to treat the disease may also be associated with mood and behavior alterations, cognitive dysfunction, fatigue, and fibromyalgia. The reader is referred to Chapters 15, 24, and 26 for a review of these concerns.

## WHY COPING IS DIFFICULT

This section lays out some of the most common everyday problems and offers a discussion of constructive approaches that can be taken to alleviate some of them. Coping with lupus calls for dealing with many different types of problems at once.

### But You Don't Look Sick!

Jane is a high-powered attorney with a prestigious big-city law firm. She usually works 50 hours a week at the office and frequently takes work home on weekends. When she began having joint aches and muscle weakness along with fatigue, her doctor diagnosed her as having systemic lupus and started her on low-dose steroids and Plaquenil. In order to keep up at work, she stopped dating and going out with friends. No one at work suspected she was ill, and no one was told. On weekends, Jane would stay in bed, barely able to move, so that she could make it to work the next week. Dr. Jones told

Jane that the drugs would take months to make any substantial difference in how she felt and that she needed to get extra rest and take care of herself. Jane, however, was afraid of losing her position and did not have time to be sick. Jane surprised herself when she finally broke down and confided in friends, who were supportive and told her to follow her doctor's advice.

For better or worse, most newly diagnosed lupus patients look perfectly healthy. Only 10 percent of patients with SLE develop a deforming arthritis, and most rashes (and paleness) can be hidden with makeup. Steroids take many weeks to puff up the face or alter one's appearance. These delays are fine if you don't want anyone to know what's going on, but it discourages the support and empathy that people who care about you might provide. Ignoring the disease can lead to a subconscious withdrawal from your vocational and social interests, mask your need for rest (as it did in Jane's case), and set the stage for disaster later on.

### My Doctor Is Not Listening

In the beginning, most patients experience some relief after being diagnosed with SLE. Some have exclaimed "I'm not crazy after all!" A honeymoon period takes place where the diagnosing doctor can do no wrong, followed by a period of questioning when the patient does not feel better immediately. I schedule a counseling session with my patients when they are first diagnosed, explain all the pitfalls of the disease, and outline a treatment plan. To fail to do this will cause problems with patients later on. Often, doctors become too judgmental and difficult to approach. Sometimes we unintentionally intimidate our patients, who are afraid to tell us of serious problems they are having that may affect treatment. Patients should not be afraid and must stick up for their rights. I always respect patients who tell me, "I need 15 minutes of your time without interruption" and give an organized, well-thought-out presentation of their problems. Also, they should not be afraid to ask for a second opinion. A good doctor will never object to this. Mutual honesty and respect, a sense of understanding of life-styles and limitations on both sides, and open lines of communication are vital. A patient's relationship with his or her doctor is akin to a complex commitment; the doctor is half of the ticket to good health, and both sides have to put up with each other's idiosyncrasies.

On the other hand, patients can try not to turn off their doctors. Several years ago, a group practice of four rheumatologists rated their combined 25 active lupus patients in terms of how much they liked or disliked them, and the same 5 patients were rated as "most disliked" by all four physicians. A psychiatrist was brought in and found that the 5 disliked patients displayed more hostility, anxiety, depression, immaturity, and uncooperative behavior patterns than the other 20 lupus patients. This group also had a tendency to doctor shop; in other words, they did not stop with a second opinion—they got ten opinions in all specialties. Anxiety, depression, and even anger are normal reactive emotions to an illness like lupus. But there are times when the doctor should be trusted; mutual trust and cooperation are vital.

### Fear and Anxiety

JoAnn was told she had lupus by her doctor, who seemed at the time very busy and distracted with other things. She had many questions to ask, blurted a few out, and did not want to bother him further. She knew that sun exposure was bad for her but did not know how much sun she could have at any one time, so she made up her own treatment. JoAnn rarely went outside. She bought a sunscreen that was the wrong strength and did not know how or when to apply it. Whenever friends invited her out during the daytime, she declined. She became fearful about keeping up her front lawn and for the first time hired a gardener. When she turned them down for a third time, JoAnn's friends stopped asking her to the lunchtime bridge games she had enjoyed so much in the past. Becoming more fearful and anxious, she tried to explain this to her doctor, who did not seem to have time to hear her out. Family members noticed that JoAnn was jumpy and agitated. When they asked her what was wrong, she said "nothing" because she did not want to impose. Family members started decreasing contact with her. Finally, she took a SLESH course that was sponsored by the local Lupus Foundation and sought counseling, which started to turn things around. She is no longer afraid to confront her doctor.

Lupus patients have many justifiable fears. "Will I be able to work?" "Can I take care of myself?" They express fear of pain, fear of disease flareups, fear of death or disfigurement, fear of drugs and their side effects. This, in turn, creates anxieties that are capable of inhibiting normal social functioning and serve to promote further isolation. "What if nobody believes me?" "Am I over- or underdoing it?" If a patient is not open and honest with friends, overanxious behavior will turn them off. Recognizing these fears and concerns and learning to deal with them is important for all lupus patients, who should express their concerns to trusting friends, family, or health-care professionals or take the SLESH course, as JoAnn did.

### Anger

After Linda was accepted to medical school, she was diagnosed with lupus nephritis. She was told that she would need high-dose steroids for months, and that these would alter mood and behavior as well as her ability to concentrate. She also had to take chemotherapy, which would make her quite sick every month. She decided to defer schooling for a year, only to learn that the school would have rejected her if it had known about her lupus. Linda became angry. It was difficult to control her emotions on 60 milligrams of prednisone a day. She was cruel to her mother, who had taken an extra job to help support her through college and no longer found time to help her younger brother with his science homework. While shopping at a supermarket, Linda, after waiting for 10 minutes on a checkout line, got into a confrontation with the store manager. When her brother got an F in science, she realized that her anger was misplaced, and, with the support of her doctor, began to channel her energies constructively toward beating the disease and pushing it into remission within a year. Linda is now a second-year medical student and is thinking of going into rheumatology.

If you have lupus, you have every right to be angry. There are several levels of anger. Rage is a violent, uncontrollable anger. Hostility, resentment, and indignation are somewhat less intense. Yes, it's annoying if you have to give up enjoyable activities, especially in the outdoors. It is also frustrating and upsetting if the treatment that will make you better won't work for several months, if you have to have surgery, or if you can no longer think about having children. It is hard for others to see how much you hurt. But don't let anger bottle up inside you. Stress has been known to cause flareups of fibromyalgia and can possibly aggravate your lupus. Anger takes up a lot of precious energy and turns off others who care about you. Try to channel your energy into productive work. Ask yourself why you are angry. How can you detect it? Can you ignore it? Step into your friend's or employer's shoes. How do they feel about what is going on? Think of how you can prevent yourself from getting angry and how to relieve anger when it builds.

## Guilt

Her mother had lupus and Nancy always knew there was a chance she would get it. But when it happened, the diagnosis seemed to drop out of the clouds. Since she had been laid off, Nancy was now convinced that the loss of her job was due to her poor performance and not the economy, even though she had good job performance ratings. Convinced that a higher authority was punishing her for being lazy, Nancy would sulk around the house and blame herself for everything that could possibly go wrong. Her mother made this worse by overcompensating and saying that because she, the mother, carried lupus genes, it was all *her* fault. Nothing constructive was accomplished in the household for months until Nancy's husband told her that he would leave if she didn't get some counseling that wouid help her develop a more positive attitude. She went for the counseling. Her lupus is now in remission and she has a new job.

Guilt is the sense that you did something wrong or that you blame yourself for something over which you had no control. Many patients inflict guilt upon themselves. ''What have I done to my children, since they have a chance of getting lupus?'' or ''Since I did not feel well enough to participate in the garden club, does this mean that I am in a weakened position so that others can manipulate me?'' Don't say ''I should have done this instead.'' Block that emotion as best you can. Guilt is self-defeating, and guilt can be a ''self-fulfilling prophecy.'' Have a positive attitude and be ready to modify your thoughts and behaviors. Counseling, such as Nancy had, can help you acquire the tools to maintain positive attitudes.

## Pain

Pain should be viewed as a sensation that is natural, inevitable, and tolerable but one that must be controlled. Most lupus patients have physical pain, but it is usually relatively mild. Only a small percentage end up attending pain centers or taking narcotic analgesics. Fear of pain, however, is a major problem. Most joint and muscle pain is inflammatory in nature and managed with anti-inflammatory medicines, not pain pills. Inappropriate reactions to pain, particularly fear of exercise and mobilization, can waste or atrophy the muscles and make joints immobile. This further promotes social isolation. Patients can help themselves deal with pain. For example, biofeedback helps to control the heart rate, blood pressure, skin temperature, and muscle tone. It can alleviate pain from headaches, spasm, and Raynaud's phenomenon. Guided imagery and meditation promote relaxation and decrease muscle spasms. Acupuncture, especially electronic acupuncture (acupressure), numbs nerve fibers and decreases noninflammatory pain.

## Stress and Trauma

Stress is a force to which the body responds. There are good forms of stress—those that energize the body—as well as types of stress that alter the immune system. Can stress cause lupus? Several years ago, members of lupus associations were asked to fill out questionnaires that included queries about how they thought their lupus had started. Between 10 and 15 percent replied that they thought that emotional stress or physical trauma brought on their disease. However, when we looked at it the other way around—examining doctors' records to see whether patients actually *were* under unusual stress (defined on a ''Life Events Inventory'' as a death of a loved one, divorce, or loss of a job) or had a serious injury shortly before being diagnosed—this percentage fell into the 1 to 3 percent range, which is not statistically significant. In other words, *it is unlikely that stress or trauma causes lupus.*

Can stress aggravate lupus? *It is very clear that stress and trauma can cause a preexisting lupus to flare up.* A large body of evidence has shown that certain animals with autoimmune disease have a defective CRH (corticotropin releasing hormone—which is made in the brain's hypothalamus) neuron, which accelerates inflammation under stressful circumstances. Not all stressed lupus patients experience flareups, and many patients have flareups when they are not stressed. Nevertheless, I have had several patients with chronic, stable, mild lupus whose disease spread to other organs or who had severe inflammatory joint reactions after a motor vehicle accident or severe emotional trauma.

Issues of stress and trauma, particularly when problems arise after an accident, can become subject to legal as well as medical definitions. For medicolegal purposes, these reactions should occur within 60 days of the incident. In state worker's compensation laws, we find the term "continuous trauma," which refers to ongoing, persistent stress or harassment that causes increased disease activity. Also included in this terminology is the tendency of a traumatic incident or continuous trauma to "light up," "accelerate," or "aggravate" lupus. There have been several cases in which patients with early nondiagnosed but symptomatic (and medically documented) prelupus who had their disease "turned on." Emotional stress can alter the immune system. Studies performed on healthy individuals who lose loved ones show that their immune functioning is altered during the bereavement period. Our "head bone" is connected to our T cells, which, in turn, affects lupus.

## Depression

Even though Judith was told her lupus was mild, things didn't seem right. Always a perfectionist who planned each activity and goal carefully, she felt that something else had to be wrong. After all, Judith could not sleep at night, she tossed and turned even though she had a new bed. She was no longer asked out on dates. She seemed distracted and always complained of headaches and cramping, and she was sure that her pulse was too fast. She no longer enjoyed watching a good movie and forgot how to tell jokes or laugh. Judith was convinced that a serious medical illness was eluding her doctors, so she went to the Mayo Clinic and the Cleveland Clinic only to be told that her local physician knew what she was doing. After joining a patient support group sponsored by the lupus society, Judith met other people with lupus who proved very insightful and pointed out each other's problems. After the third session, the group unanimously told her that she was depressed and that her symptoms were part of this reaction. The group leader, a psychologist, recommended a psychiatrist, who started her on an antidepressant. Within 4 months, she was off all medication and leading a normal life.

Depression is the most common coping problem in lupus. Lupus itself and some of the medications used to treat the disease can induce a clinical chemical depression. Additionally, lupus patients can develop a reactive depression: they are upset that they have the disease. In the simplest terms, depression is a feeling of helplessness and hopelessness. It is characterized by spells of crying, loss of appetite (or increased appetite), nonrestful sleep, loss of self-esteem, inability to concentrate, decreased social interests, indecision, and loss of interest in the outside world. Physiologic signs and symptoms such as headache, palpitations, loss of sexual drive, indigestion, and cramping may go along with depression. Depression affects your body, mood, relationships, and physical activities.

Some suggestions of ways to cope with depression are detailed in the next section. First, your doctor should verify that the depression is not due to central nervous system lupus or medications you are taking. For example, steroids both induce and help depression. Chemical or biological depression is treated differently from reactive or psychological depression.

## HOW TO COPE BETTER

Depression, fear, and anxiety are the most common reactions noted in SLE patients. How should they deal with these emotions? How can they cope better? Adequate coping requires taking action, and several good approaches are available.

### Goals and Attitudes

If you are a lupus patient, your first step is to define and try to assess what is bothering you. Professional help may be needed to do this. Then you deal with these concerns by listing realistic goals and expectations. Develop a method for problem solving: problems can be eliminated, circumvented, modified, or worked out. Use all your available resources: financial, personal, and intellectual. Be willing to reassess if you cannot meet your goals or expectations. Don't try to change others; try to change yourself. Set realistic goals for yourself to improve your quality of life. Pace yourself and allow for periods of activity that alternate with periods of rest. Ask yourself what hopes you have and—if they are unrealistic—try to replace them with other hopes. Try to balance a loss with a gain. How can you improve your spiritual well-being? Learn to relax, learn to rest and to exercise. If you don't have a sense of humor or if it is suppressed, discover laughter. Laughter is an excellent tonic for the body. Give affection to others and you'll receive it back. Learn to share yourself. Don't worry about tomorrow and focus on what you can do today. What kinds of things do you like to do and how can you do them? Exchange negative thoughts for positive thoughts and reinforce them. Socialize. All these things help to make you feel better about yourself and to conquer depression.

## Marriage, Family, and Sexuality

Darleen and George were happily married for 5 years when Darleen was diagnosed with SLE. George had grown up with learning difficulties and had had limited educational opportunities. Darleen tried to tell him what lupus was, but he didn't seem to pay attention. When Darleen was put on steroids and gained 20 pounds, George made fun of her appearance. One night her joints were so swollen that she couldn't even get into the car to go to George's friends' house for dinner. George said that her joints looked OK to him and started yelling at her. Over the next few months, George started drinking heavily and lost interest in sex. Darleen was scared to talk to him, and one day he just didn't come home.

Unfortunately, reports suggest that within 5 years of the diagnosis of lupus, nearly half of married women are divorced. This results from many of the emotional changes discussed above and a coping problem on the spouse's part. (''What do you mean you can't go out with me tonight? You look fine!'') When women complain of difficulty in keeping up with household chores, or workplace demands, or responsibilities to their children, relationships become precarious. After they have been diagnosed, I ask lupus patients to bring their boyfriends or husbands to a counseling session. They shouldn't feel that they are ''out of the loop'' or that the doctor may be hiding things from them. If possible, spouses should be included in any decisions.

Spouses should know that steroids can alter appearance, mood, and behavior and that family responsibilities might have to be shifted for a time. Parents may ignore problems, smother the patient, or act somewhere in between and be appropriately supportive. It is up to the patient to decide what role they should be assigned, if any, as part of the recovery plan.

Surprisingly, very few of the reasons for divorce among patients with lupus have anything to do with sexuality. A detailed survey showed that only 4 percent of women with SLE had major problems with sexuality. Most of these cases dealt with a dry vagina from Sjögren's syndrome (also causing dry eyes, dry mouth, and arthritis) that is difficult to lubricate and can cause painful intercourse. Other cases involved women who understandably complained of being too tired to participate in sex. Destructive hip changes from arthritis or avascular necrosis also make lovemaking difficult, but they are easily resolved with creative sexual positions and/or corrective surgery. Divorce or separation arises from not being frank with a loved one, altered expectations, lack of knowledge about lupus and how it can affect mood and behavior, and from husbands' reactions to learning that their wives cannot bear children—which of course does not apply to all women with lupus. (See Chapter 30, "Can a Woman with Lupus Have a Baby?") Keep all communication channels and support systems open!

## Support Groups, Self-Help, and Counseling

The Arthritis Foundation and lupus associations provide self-help groups supervised by trained professionals like psychologists or social workers. They provide intrinsic support, disease education, emotional warmth, means for friendly communication, and closeness with others. Working together, they can help a patient reverse a negative self-image and self-defeating attitudes through a supportive atmosphere. This can lead to more hope, improved self-esteem, a redirection of energy and a sense that one is not alone.

Sometimes, one-on-one psychologic counseling is advisable. On occasion, medication is required, and this may be prescribed by a psychiatrist. Psychiatrists are medical doctors, all of whom are taught about lupus and autoimmunity in medical school. Psychotherapists and rheumatologists should work together as teams. Tricyclic antidepressants such as Sinequan, Pamelor, Elavil, and Norpramin promote restful sleep, raise pain thresholds, relax muscles, and improve mood. Serotonin reuptake inhibitors such as Prozac, Effexor, Zoloft, or Paxil may be used with or without tricyclics. Taken in the morning, they often increase energy levels, promote weight reduction, relieve depression, and diminish obsessive-compulsive tendencies. Additional goals of psychological intervention are to increase self-control, patience, tolerance, flexibility, and creativity.

## Disability

Can lupus patients work? Several large surveys show that 60 to 70 percent of lupus patients are able to work. Social security disability income and Medicare hospital and physician insurance are available to lupus patients who fulfill the ACR criteria for lupus (Chapter 2), with frequent exacerbations of the disease in the critical organs. Benefits may begin if you have not worked for 2 years because of the disease. The overwhelming majority of patients with non-organ-threatening disease are able to work full time if they are allowed rest periods, can avoid the sun, and do not participate in overly heavy physical labor that would require constant use of the joints. Most states provide vocational rehabilitation to individuals who can work but cannot perform the type of work they used to do. As one rheumatologist put it, lupus patients are not disabled, they are "differently abled." Patients with SLE work best when they can work at their own pace and control their environment. For example, professionals with SLE often thrive if they are doctors, lawyers, accountants, or in business for themselves and can set their own hours. Vocational counselors often steer individuals with lupus without a college diploma or professional training to positions as travel agents or realtors or to home-based mail-order or computer-related businesses. This provides them the freedom to work when they feel up to it. Sometimes, a diagnosis of lupus gives one a good opportunity to return to school and obtain additional education.

## Children and Adolescents

Children with lupus need to be treated like any other children. Although there are restrictions on sun-related and certain other activities, *they should not suffer because of their parents' guilt and they should not be overprotected.* Many children instinctively deny their lupus; this is OK if they take their medicine and follow the usual precautions. Be matter of fact with them in discussing the disease.

On the other hand, children of mothers with SLE are often unusually astute and aware of their mothers' problems. Most of my female patients with young children have been asked at least once, ''Mommy, are you going to die?'' Be honest with your children and don't hide important things from them. They are bound to find out. Couch whatever you tell them in hopeful and positive terms. They can be a source of pride and joy and deserve to be part of your support system.

*Teenagers with lupus present special problems.* They are concerned with hair loss, rashes, and fatigue, which prevents them from participating in social activities, and with sun exposure (Chapter 24). They want to know about pregnancy. Adolescents need a certain degree of independence and responsibility; they need friends and often try to be away from parents, whom they perceive as embarrassing to them. Steroids alter appearance, hair growth, mood, and behavior, which affects dating, jobs, and school life. Compliance is a major problem and the consequences of not taking the prescribed medication or less of it should be emphasized.

Joelle is a 16-year-old girl with multisystem active lupus. It had been under good control until she fell in with a new crowd that went to the beach every day and stayed out very late at night. It was summertime and her rashes started getting worse. This flared her joints and started causing low-grade fevers. Because she put on a partially adequate sunscreen, Joelle did not think that sunbathing was wrong and told her mother she was going out with friends. Instead of taking the 20 milligrams of prednisone prescribed, she took only 10 milligrams because she did not want her face to look puffy or to develop facial hair. Ultimately, her doctor noticed that her anemia was so severe that she was on the verge of needing a transfusion. She was therefore hospitalized. A few days of intravenous high-dose steroids stabilized her and the medical interns and residents on the case spent extra time with Joelle giving her emotional support. Her family was called to Joelle's bedside and the doctors discussed with them the importance of compliance, sun avoidance, and the dangers of not adequately treating the disease. A support system was devised to prevent further problems.

Approaching the adolescent requires unique solutions. First, talk to school personnel and see if they will work with the child. See if there is any way concerned classmates and educators can be informed about the disease and the special considerations that may apply. Also, are there important people in the teenager's life who are role models, such as trainers, coaches, teachers, clergy, or extracurricular activity instructions that can be brought into the loop? Finally, try to direct the teen's energies into constructive hobbies or interests or safe projects.

### Avoid Unproven Remedies

There are times when lupus patients can feel desperate, and medical quackery is a multibillion-dollar-a-year industry whose claims seem tempting at certain moments. Any promotion that offers a cure for lupus should be suspect. Don't believe testimonials if the article has not appeared in the peer-reviewed medical literature. Be careful of treatments that are very expensive, such as chelation therapy, fetal animal hormone extracts, or monthly gamma globulin infusions. Some ''natural vitamin and mineral'' products contain alfalfa sprouts or other suspect chemicals that can aggravate lupus. L-tryptophan was an over-the-counter ''natural'' supplement whose metabolites induced scleroderma in 2000 people and killed 50 of them before it was removed from the market in 1989. Mexican border clinics that promise cures treat lupus with steroids combined with a dangerous NSAID called phenylbutazone, which is no longer available in the United States, since it can cause leukemia and bone marrow shutdown. Some Chinese herbal remedies contain sulfa derivatives and other substances that trigger allergic reactions in most lupus patients. Consult a physician before using any nonprescription drugs or potions. One of the first things I learned in medical school was ''Do the patient no harm.'' And *you* should remember ''Caveat emptor!''

## Summing Up

If you have lupus, you have at least half a million Americans for company. Take hold of yourself and don't become overly upset. If there's no organ-threatening disease, you have a normal life expectancy; even if there *is* such disease, you will still live a long, long time. The disease will not go away. Although its course waxes and wanes, it is a permanent, chronic condition. It's OK to be angry, frightened, guilty, anxious, and depressed at first, but these emotions can be overcome. The sooner you develop adequate coping mechanisms, build up a good support system among family and friends, and start to work on a positive and realistic life-style, the better you will feel physically and mentally. Who knows, it may even make the lupus better!

# 26
# *Taming Inflammation: Anti-Inflammatory Therapies*

Even though most patients with SLE don't like it, more than 90 percent take medication. Of the many drug therapies for lupus, and among the most prescribed and accepted are what doctors call "the anti-inflammatories." These remedies include *aspirin products* and other *nonsteroidal anti-inflammatory drugs,* or NSAIDs. A recent survey has documented that at any given time, 80 percent of all SLE patients are taking one of these agents on a regular or intermittent basis. This chapter will endeavor to help lupus patients negotiate the tricky ins and outs of NSAIDs in an effort to maximize potential benefits while minimizing toxicity.

## ASPIRIN AND ITS FIRST COUSINS ONCE REMOVED

The fever-lowering property of willow bark (*Salix alba*) was known to the ancients and used by Hippocrates, Galen, and Pliny. The active salicylate ingredient was isolated in France in 1827, and acetylsalicylic acid (aspirin) specifically identified in 1899 in Germany. When Dr. Marian Ropes founded the first lupus clinic in Boston in 1932, aspirin was the only real medication she had to work with. For reasons that are still unclear, only one well-designed study has ever assessed the efficacy of aspirin in SLE. Fortunately, it was a definitive one. In 1980, the National Institutes of Health proved beyond the shadow of a doubt that aspirin was helpful in SLE.

### What Does Aspirin Do and Not Do in SLE?

The salicylate class of medicines, of which aspirin is the most important, ameliorates lupus by lowering fevers, treating headaches, diminishing joint or muscle aching and inflammation, and decreasing both serositis (pleurisy, pericarditis, peritonitis) and malaise. It has no effect on skin, heart, lung, kidney, liver, central nervous system, or blood involvement of the disease. Salicylates do not modify the disease; that is, they do not put lupus into remission and do little about the underlying immune process. Aspirin relieves lupus through the inhibition of prostaglandin, a chemical that promotes inflammation and pain in arthritic disorders.

## The Many Faces of Aspirin

Aspirin can be administered in many ways. Along with the tablet form, it is available as a suppository, a liquid suspension (Arthropan), or a salve (e.g., Ben-Gay). Most people, however, take it orally. Aspirin can be buffered to diminish stomach irritation (e.g., Bufferin); these preparations work quickly (in 10 to 15 minutes) but are out of the system in 3 to 4 hours. Two regular or unbuffered aspirins (325 milligrams) are effective in less than an hour and work for 4 to 6 hours. Patients with headaches often benefit when caffeine is added to aspirin (e.g., Excedrin). The anti-inflammatory dose of aspirin depends on the individual and how that person's liver and kidney will handle the drug. An active rheumatoid-like arthritis responds to 325-milligram tablets taken three or four at a time four to five times a day. Since this involves popping a lot of pills, rheumatologists frequently prescribe high-dose timed-release preparations (e.g., Zorprin), which last 8 to 12 hours and need to be taken only two or three times a day. Salicylate blood levels are monitored in some patients to see if they are taking enough aspirin to enjoy its maximal benefits. Too much aspirin can cause ringing in the ears; an old saw claims that the correct dose of aspirin for rheumatoid-type arthritis is one tablet a day less than what causes the ears to ring.

Low-dose aspirin (baby aspirin, 81 milligrams) may be prophylactic in lupus patients with antiphospholipid antibodies (Chapter 21). Why low-dose and not regular aspirin? Low-dose salicylates inhibit a chemical that promotes clotting; regular aspirin also inhibits this chemical but additionally inhibits a second chemical that counteracts this. Taking 1 grain (81 milligrams) of aspirin daily may decrease the risk of strokes or miscarriages due to the lupus anticoagulant.

Salicylates other than aspirin are slightly modified. Some of these should not induce bleeding, rarely upset the stomach, and cannot cause ulcers. These prescription sodium or magnesium salicylates (Magan, Trilisate, Disalcid) are excellent anti-inflammatory preparations for patients who have had an ulcer or look bruised all the time. Unfortunately, they are much weaker than regular aspirin and are often ineffective.

## What Are NSAIDs and Why Are They Used in Lupus?

Pharmaceutical companies began their search for a better aspirin in the 1940s and in 1952 came up with *butazolidin* (*Phenylbutazone*). However, this is no longer marketed in the United States. It was followed by *indomethacin* (*Indocin*) in 1965 and *ibuprofen* (*Motrin*) in 1974. Over the last 20 years, numerous additional preparations have been introduced: they are listed in Table 12 and discussed below. All of these preparations are more potent than aspirin, so that fewer pills are required to achieve the same effect. However, very few of these drugs are superior to aspirin: they have more side effects and are all more expensive.

**Table 12.** *Major Nonsteroidal Anti-Inflammatory Drugs (NSAIDs)*

| |
|---|
| Salicylates |
| Aspirin |
| Sodium salicylates (Trilisate, Disalcid) |
| Diflusinal (Dolobid) |
| Magnesium salicylate (Magan, Doan's) |
| Proprionic acid derivatives |
| Oxaprozin—Daypro |
| Naproxen—Naprosyn, Anaprox, Aleve |
| Flurbiprofen—Ansaid |
| Ibuprofen—Motrin, Advil |
| Ketoprofen—Orudis, Oruvail |
| Fenoprofen—Nalfon |
| Acetic acid derivatives |
| Sulindac—Clinoril |
| Diclofenac—Voltaren, Cataflam |
| Tolmetin—Tolectin |
| Indomethacin—Indocin |
| Others |
| Etodolac—Lodine |
| Ketrolac—Toradol |
| Piroxicam—Feldene |
| Nabumetone—Relafen |
| Meclofenamates—Meclomen, Ponstel |

All the NSAIDs, including aspirin, function similarly; their most important effect is inhibiting prostaglandin. But very few drug trials using these agents in SLE have been published. The manufacturers of NSAIDs have been careful not to promote the use of these drugs in SLE, since this would be ill advised if the disease should affect the kidney or liver. (That is a possibility in up to half of all patients.) In fact, no NSAIDs are FDA-approved for use in SLE. In spite of this, as noted, 80 percent of all lupus patients take NSAIDs on a regular or intermittent basis. Some patients take NSAIDs on a daily, regular basis while others only use them when they have pain or inflammation. The reasons for this are that the NSAIDs are effective in relieving fevers, headaches, muscle aches, malaise, arthritis, and serositis associated with SLE. All NSAIDs have widely varying pain-relieving properties.

### Which NSAID Should Be Used?

The ideal NSAID is inexpensive, safe, and effective; however, none of them fits this billing exactly. Below, we examine each NSAID and explain the pros and cons for using each. *Flurbiprofen* (*Ansaid*) may uniquely counteract osteoporosis and mineralize bone. It is of moderate potency and rarely induces changes in the patient's mental status. *Naproxen* (*Naprosyn*) is the best-selling NSAID in the United States. Taken twice a day and available as a liquid suspension, it is highly potent, but the possibility of gastritis requires careful monitoring. A salt of naproxen called *Anaprox* (*available over the counter as Aleve*) takes 15 minutes to work (as opposed to 2 hours for naproxen) and is used as an analgesic for headaches or menstrual cramps, for example. *Oxaprozin* (*Daypro*) is a highly effective once-a-day "timed-release Naprosyn." Doses of these drugs should be decreased in older patients, and they can all be toxic to the stomach. Of all the NSAIDs, *fenoprofen* (*Nalfon*) causes the most salt retention and is among those that are most toxic to the kidneys. This weak preparation must be taken four times daily and has no place in managing SLE. *Ketoprofen* (*Orudis*) is a moderately potent preparation used three times a day. Its major advantage is a rapid onset of action (10 minutes) and highly effective analgesic effects. Therefore I prescribe ketoprofen when a quick response is mandated for intermittent pain. It is also available as a 24-hour timed-release preparation (*Oruvail*). *Ibuprofen* (*Motrin, Advil, Nuprin, etc.*) has flexibility of dosing (200-, 400-, 600-, and 800-milligram strengths) and is available without a prescription. This weak agent is taken four times daily if used regularly. (Chapter 9 reviews some of the problems with these drugs, especially the central nervous system reactions observed with ibuprofen.)

*Diclofenac* (*Voltaren*) is the world's best-selling NSAID for arthritis in general. An extremely effective analgesic of moderate potency, it is coated to protect the stomach and has mild anti-inflammatory properties, as does its close relative *Cataflam*. However, liver function must be monitored every 3 months. *Piroxicam* (*Feldene*) is a highly potent agent that has the advantage of being taken once a day. But many lupus patients have difficulty tolerating its side effects: it can enhance sun sensitivity. Piroxicam can also be hard on the stomach. *Sulindac* (*Clinoril*), *indomethacin* (*Indocin*), and *tolmetin* (*Tolectin*) are associated with a 10 to 15 percent incidence of central nervous system complications (e.g., headache, mental clouding) not observed with any other NSAIDs. Sulindac is moderately potent, well tolerated, and taken twice a day. Indomethacin is the strongest NSAID on the market and is also available as a suppository and as a 24-hour timed-release oral preparation. Intended for short-term use, it carries the greatest risk of toxicity to the liver, stomach, and kidneys with continuous use. Tolmetin is a moderately potent preparation that infrequently prolongs bleeding times and thus may be used in dialysis patients and prior to surgery.

*Meclofenamate* (*Meclomen*) is an expensive, weak agent that induces diarrhea in 30 percent of patients but may help psoriasis. *Etodolac* (*Lodine*) is a mild, very well tolerated NSAID that has an excellent kidney, stomach, and liver safety profile. The strongest NSAID analgesic preparation is *Ketorolac* (*Toradol*), which is available orally or by injection. Approved only for short-term or occasional use, it has minimal anti-inflammatory properties. Finally, *nabumetone*

(*Relafen*) is a once-a-day medication of moderate potency that causes ulcers less often than any of the preparations mentioned above and has an excellent liver and kidney safety profile. Some patients report that this agent increases their energy levels.

## Why Must We Be Careful When Using NSAIDs?

The NSAIDs can be extremely effective. Their use can mark the difference between holding a job and being on disability. They can greatly improve one's quality of life. However, numerous caveats apply when using these drugs. First of all, *all patients taking NSAIDs on an ongoing basis should have complete blood counts as well as liver and kidney blood chemistries every 3 to 4 months.* An increase in liver enzymes to greater than 2.5 times normal mandates discontinuation of the NSAID so as to protect the body from liver failure. Similarly, patients with lupus nephritis probably should not take NSAIDs unless it is for a specific circumstance and under close medical supervision for a limited period (e.g., to treat acute gout or bursitis) and renal function is carefully monitored. All NSAIDs have the capacity to induce kidney failure. In patients with normal liver and kidney function, the risk of developing altered function ranges from 0.1 to 10 percent, depending on the agent employed. Fortunately, most patients have no symptoms, and these abnormalities are almost always reversed when they stop taking the drug.

Most importantly, NSAIDs commonly induce erosions in the stomach, and this may lead to bleeding ulcers. Nationwide, 3 percent of all patients taking NSAIDs on a regular basis develop bleeding ulcers. A low hemoglobin or hematocrit (part of the complete blood count), black stool, or a positive Hemoccult test (stool smear for blood) are often the first signs of trouble. Tobacco and alcohol abuse increase the risk of ulcers. The use of $H_2$ blockers (Zantac, Pepcid, Axid, Tagamet), antacids, or sulcrafate (Carafate) relieves dyspepsia and heartburn associated with NSAIDs (seen in 20 to 30 percent) but do *not* prevent the development of ulcers. I advise patients who need to be on NSAIDs and have had gastric erosions or ulcers to add omeprazole (Prilosec) or misoprostol (Cytotec) to their regimen. These drugs *prevent* ulcers from developing.

Other adverse reactions can result from NSAIDs. These include bloating and fluid retention, easy bruisability, diarrhea, ringing in the ears, headaches, provocation of allergy or asthma attacks, and rashes. Because NSAIDs prolong bleeding times, they should be discontinued at least a week before any surgery so as to minimize bleeding risks. Doctors may try as many as ten agents before they find one that works and has no side effects. The choice of NSAID preparations is a highly individual one, and what works for one person may not work for another.

## WHAT ABOUT ACETAMINOPHEN (TYLENOL)?

Acetaminophen is an analgesic medication that modestly decreases pain (e.g., headache) and lowers fevers. It has *no* anti-inflammatory effects, so it is usually not recommended in lupus cases. Rheumatologists occasionally come across patients who cannot take any NSAIDs; therefore they may reluctantly advise acetaminophen. The only other time it is used is to relieve pain for a week prior to surgery, when NSAIDs should not be used (because they prolong bleeding times), and when a pregnant woman has other nonlupus illnesses causing pain or discomfort such as headaches, fevers, or back sprain.

### Summing Up

Salicylates have been used for more than 100 years and other nonsteroidals for more than 40 years. They are effective in treating fevers, headaches, arthritis, arthralgia, muscle aches, and serositis. None of these agents have any disease-modifying properties in lupus and have no place in serious SLE without disease-modifying drugs. The choice of a particular drug depends on the medical history; several agents are usually tried before one is found that is effective and safe. Call your doctor if you encounter any problems with these drugs. All the preparations listed in this section require regular blood testing and a rectal examination with a stool smear for blood at least once a year. With appropriate monitoring, NSAIDs can be given concurrently with other anti-lupus medications.

# 27

# *Big Guns and Magic Bullets: Disease-Modifying Drugs*

Up to this point, we've covered nonmedical aspects of therapy as well as the over-the-counter basics and prescription—NSAIDs—which help control symptoms and pain. Now we move on to other prescription drugs. Over 90 percent of all lupus patients will be given one of these agents at some time in the course of their disease. These drugs are termed "disease-modifying" because they can actually alter the course of lupus. Rheumatologists have nicknames for them such as DMARDs (disease-modifying anti-rheumatic drugs), among others. The DMARDs for lupus fall into three general categories: antimalarials, steroids, and chemotherapy.

## WHAT DOES LUPUS HAVE TO DO WITH MALARIA?

When a rheumatologist mentions to a patient that she might benefit from antimalarial therapy, the usual response is one of confusion. The relationship was discovered by accident. The first use of antimalarials dates back to 1630, when the Countess Anna del Chinchon, wife of the viceroy of Peru, fell seriously ill with a high fever. In desperation, the family allowed local Indians to treat her with bark extracts from a "fever tree," and she made a miraculous recovery. The cinchona tree was named after the grateful countess, who proceeded to corner the market and arrange its exportation to Spain. Years later, quinine was found to be the active ingredient of cinchona bark. In addition to its ability to diminsh fever, it was noted to have anti-inflammatory and anti-infectious properties, including beneficial activity against malaria. When the Japanese conquered Java during World War II, they effectively prevented the Allies from obtaining natural quinine. Faced with a serious threat of malaria among their troops in the south-western Pacific theater, the Allies bootlegged an artificial quinine recipe from the notorious German I. G. Farbenindustrie combine (which made the gas chambers used in the concentration camps) and called it Atabrine. In 1943, the U.S. Surgeon General declared Atabrine to be the drug of choice to prevent malaria, and 4 million American, Canadian, Australian, New Zealand, and British soldiers took it daily for 3 years. There were anecdotal reports that soldiers with rheumatoid arthritis or lupus experienced improvement in rashes and joint symptoms when they took Atabrine; this led the British to conduct a study on its efficacy in lupus.

The findings were published in the English journal *Lancet* in 1951. Antimalarials have been with rheumatologists ever since. Efforts to improve upon Atabrine led to the introduction of chloroquine (Aralen) in 1953 and hydroxychloroquine (Plaquenil) in 1955.

## How Do Antimalarials Work?

Antimalarials have remarkably different and often divergent actions. They block ultraviolet light from damaging skin; have an anti-inflammatory effect similar to that of the NSAIDs; lower cholesterol levels by 15 to 20 percent; inhibit clotting; block cytokines, which promote inflammation; and, most importantly, alter the acid-base balance of the cells, which limits their ability to process antigens. If antigens are allowed to be processed, this would lead to the creation of unnecessary antibodies. Unlike steroids and chemotherapies, antimalarials do not lower blood counts or make patients more susceptible to infection. They are mild by nature, and altering the acid-base balance of cells takes time before a significant effect is noted. Plaquenil, for example, has an onset of action of 2 to 3 months, but its efficacy does not peak for several more months. Most patients are unaware of its benefits for 4 to 6 months. These agents have no place in the management of organ-threatening SLE.

## What Symptoms and Signs of Lupus Respond Best to Antimalarials?

In patients with non-organ-threatening SLE, antimalarials benefit several systems. The skin is helped by the interruption of ultraviolet light and anti-inflammatory actions. Discoid lesions, redness, mouth ulcers, and hair loss improve in 90 percent of these patients. Joint manifestations also diminish as swelling and aching decrease. Over time, inflammation of the pleura and pericardium lessens, as do constitutional symptoms of fatigue and cognitive dysfunction. Antimalarials are mild cortical stimulants; they are similar to mild amphetamines in the sense that they can give patients energy. This is particularly true with Atabrine and to a lesser extent Aralen or Plaquenil. Antimalarials are also being studied for possible use in patients with Sjögren's syndrome (dry eyes, dry mouth, and arthritis), steroid-induced high cholesterol, and antiphospholipid antibodies that pose a risk for blood clots.

In 1991, a group of Canadian rheumatologists studied 47 patients whose mild, non-organ-threatening lupus was under good control with Plaquenil. Half of the patients were randomized to receive a placebo (sugar pill) and half continued their Plaquenil. This was a double-blind study—that is, the patients had no idea which drug they were being given, nor did the reseachers. The Plaquenil group had many fewer disease flareups and organ-threatening complications over a 6-month period. This study confirmed previous suggestions that the institution of antimalarial therapy early in the disease course decreases the risk that lupus will spread to critical organs.

## What Antimalarial Should Be Taken, in What Dose?

The only antimalarial currently approved by the FDA and promoted specifcally for lupus is Plaquenil. The cost of a year's therapy is about $500. Rheumatologists vary widely in their use of this drug. I have my newly diagnosed lupus patients without critical organ disease take Plaquenil for at least 2 years. The usual dose is 400 milligrams (two tablets) once daily. During the third year of therapy, I reduce the dose to 200 milligrams a day (one tablet). Most of my patients are 80 to 90 percent better after 2 to 3 years. About half of them are able to discontinue the drug after 3 to 4 years and half need additional treatment. Sometimes, acute flareups necessitate higher doses of the drug; the FDA allows up to 800 milligrams of Plaquenil to be given daily for short periods of time. Some lupus patients might require longer or different Plaquenil regimens. These include patients with Sjögren's syndrome. Dr. Robert Fox at the Scripps Institute in La Jolla, California, has shown that it takes 2 to 3 years of therapy before dry eyes and dry mouth show clinical improvement. Patients with critical organ disease or serious joint inflammation who require steroid therapy might benefit from the concurrent use of Plaquenil as long as they are on corticosteroids, since the antimalarial is steroid sparing (allows lower doses of steroids to be given), lowers cholesterol raised by steroids, and may decrease the tendency of steroids to promote clots.

What about the patient who has only a partial response to Plaquenil? After 6 months of treatment, many patients tell me that they are 50 percent better but still have rashes or joint swelling. In these cases I first make sure they have had an adequate trial of all appropriate NSAIDs that can be given along with antimalarials. If this is the case, we can add methotrexate (discussed below), low doses of oral steroids (discussed immediately below), or another antimalarial.

What other antimalarials are appropriate for lupus? *Chloroquine* (*Aralen*) was FDA-approved for lupus and rheumatoid arthritis and is still widely used outside the United States for these indications. But the manufacturer no longer recommends it for lupus because it also makes Plaquenil, which is much safer. Nevertheless, chloroquine is readily available and is a much more potent drug. It is particularly effective for skin rashes and joint inflammation, and therapeutic benefits are noticeable in a month. I prescribe chloroquine when patients with severe skin or joint lupus cannot wait 6 months for Plaquenil to work, but I switch them over to Plaquenil within 3 to 4 months to prevent serious eye toxicity. As of this writing, the production of generic chloroquine has been discontinued in the United States, but Aralen is available. *Atabrine* is still available through 2000 "compounding pharmacists" in the United States as quinacrine. These old-time pharmacists make the drug from scratch by grinding up quinacrine powder. It may again be available as a tablet in the future. Atabrine is indicated for lupus patients who cannot risk any potential damage to their eyes, those who cannot tolerate Plaquenil (it is chemically different), and those whose main complaint is profound fatigue. Atabrine is synergistic with Plaquenil or chloroquine: the drugs mix together well and their combined benefits are greater than those of both agents separately. The usual dose for Atabrine is 100 milligrams daily, although as little as 25 milligrams a day is effective. Atabrine works within a month, and after 2 years it can be tapered to 5 days a week, then 3 days a week, and finally once a week.

For those who remember my friend Ray Walston standing and dancing on a box labeled Atabrine during the soldiers' variety show in the movie *South Pacific*, it has been half-jokingly suggested that this cortical stimulant played a role in the Allies' victory in World War II.

## General Adverse Reactions and How to Handle Them

Plaquenil is generally very well tolerated. A 1992 surveillance study showed that only 8 percent of lupus patients given the drug discontinued it during the first 12 months. Approximately 10 to 15 percent of patients develop generalized complaints of aching, nervousness, headache, queasiness, or nausea. When a patient relates these problems to me, I stop Plaquenil for 72 hours and then restart it at 200 milligrams daily. The body adjusts to the drug, and within a few months the dose can be increased to 300 to 400 milligrams. Doses as low as 200 milligrams a day can be effective in lupus; they just take longer to work. Another 5 percent of patients develop a rash on Plaquenil. When this occurs, the drug must be discontinued. Chloroquine has the same side effects as Plaquenil, but they occur more often. In addition, the drug may sometimes (less than 1 percent of the time) cause hair to gray, damage muscle cells, and lower blood counts. These reactions are almost unheard of with Plaquenil. Long-term use of both drugs can cause black or blue skin pigment deposits. Atabrine produces mild abdominal cramps and diarrhea in up to 30 percent of patients; this is managed by taking Pepto-Bismol. Also, there may be a reversible yellowing of the skin in 40 percent of patients taking at least 100 milligrams daily (lower doses rarely cause skin changes). As a mild "upper," Atabrine is capable of inducing a toxic psychosis with manic behavior in 1 of 500 patients, but that disappears with its discontinuation. Blood counts should be monitored, although the risk of problems in the absence of an allergic rash is 1 per 30,000. If an allergic rash breaks out, therapy should cease immediately.

### *What about the Eyes?*

My pet peeve is that too many lupus patients have been frightened away from antimalarials by doctors who warn them about potential eye toxicity. First of all,

Atabrine, for one, does not affect the eyes in currently used doses. Second, eye toxicity caused by Plaquenil is extremely unusual. It is so rare, in fact, that the British National Health Service stopped covering routine eye examinations for its Plaquenil patients. Melanin (normal skin pigment) deposits at the back of the eye, or retina, can be promoted by antimalarials, causing blind spots (called scotomas), since we cannot see through melanin. The risk of this occurring with Plaquenil is 3 percent with 10 years of continuous use, or in 1 patient out of 30. Patients tell me all the time that after a month on the drug, they have blurred vision. They are not making it up; steroids can not only cause blurred vision but also can lead to cataracts and glaucoma. Also, many patients started on Plaquenil are also on steroids. If a lupus patient sees a retina specialist every 6 months, the deposits will be noticed before they cause any symptoms, and if the drug is stopped, the deposits will disappear in weeks. *There has never been a case of permanent eye damage from Plaquenil with normal kidney function in any patient who has had eye examinations every 6 months and who has taken the drug for less than 10 years.*

On the other hand, chloroquine can cause permanent eye damage and must be used very carefully. The risk of eye toxicity with chloroquine is 10 percent after 10 years, and the changes it causes are not necessarily reversible. All patients on chloroquine should be examined every 3 months. Additionally, cloudy vision caused by swelling of the front of the eye (called corneal edema) is common with chloroquine and rare with Plaquenil. It is usually so mild that most patients continue to take the drug, and it goes away once these agents are stopped. With our increasingly sophisticated technologies, ophthalmologists are coming up with newer methods to evaluate retinal function. Many university medical centers are trying these technologies out on lupus patients who take antimalarials. The bottom line is that many subtle abnormalities of no consequence are noted in reports, and this may serve to scare patients, unnecessarily, away from antimalarials. Often, these subtle abnormalities are not found when the tests are readministered months later. Eye changes from Plaquenil take many years to develop, and any risk of damage disappears when the drug is stopped.

## IS CORTISONE REALLY NEEDED?

Steroids are the most effective and most misunderstood treatment for lupus. They are also the most used and abused therapeutic interventions for the disease. Simply stated, if organ-threatening disease is present and steroids are not prescribed, the patient usually loses function in that organ. If mild disease activity is present, other therapeutic alternatives with or without steroids are available, but many physicians have little experience in using these alternatives and tend to overuse steroids. How did things get this way?

Let us begin our story in 1855, when Thomas Addison noticed that the adrenal gland above the kidney had physiologic functions. Even though it was soon shown that the hypothalamus and pituitary controlled some of these functions, the adrenal gland's ability to act independently on occasion excited interest in this area. Tissue from adrenal cortex extracts was broken down and analyzed. By 1942, the structures of 28 steroids from the adrenal cortex had been elucidated. Ultimately, five of them were shown to be biologically active: in other words, they did something. Enter Phillip Hench, the only rheumatologist ever to win the Nobel Prize in Medicine. As early as 1929, he had observed that patients with rheumatoid arthritis went into remission during pregnancy, and he speculated that a metabolite was responsible. As soon as Hench became aware of the ongoing research on the adrenal cortex, he speculated that these steroids might be the "metabolite" he was looking for. In 1949, Dr. Hench and his colleagues at the Mayo Clinic gave cortisone to a patient with rheumatoid arthritis, who had a dramatic response. Numerous steroid preparations became available during the 1950s, and rheumatologists eventually determined—largely as a result of trial and error—which preparations should be used and in what amounts for inflammation. Lupus patients first received cortisone in 1950, and they have been using its derivatives ever since.

## How Do Steroids Work?

Steroids are hormones that play an important role in the body regulating *physiologic* functions. Doses above and beyond what the body makes have different actions, termed *pharmacologic* functions. The pharmacologic activities of steroids include stabilizing cells so that they are less likely to become involved in inflammatory processes. Steroids block numerous chemical pathways. They also decrease the number of circulating lymphocytes—the white blood cells responsible for immunologic memory.

Within the brain lies the hypothalamus, an organ that makes a chemical called *corticotropin-releasing hormone* (*CRH*). This hormone travels downstream a short distance to the pituitary gland and induces the pituitary to make *adrenocorticotropic hormone* (*ACTH*). The ACTH then stimulates the adrenal cortex to make steroids. This network, known as the *hypothalamic-pituitary-adrenal axis,* governs how much steroid our body makes. An average adrenal gland makes the equivalent of 7.5 milligrams of prednisone a day. Through a negative feedback loop, patients taking more than 7.5 milligrams of prednisone daily prevent the release of CRH and ACTH, which stops the adrenal gland from making steroids. This is why steroids cannot be discontinued suddenly; it takes months for the hypothalamic-pituitary-adrenal axis to return to normal after steroid doses drop below 7.5 milligrams of prednisone a day. Adrenal insufficiency, also called Addison's disease (named after the discoverer of the adrenal cortex), is a potentially serious condition characterized by a critical shortage of steroids in the body.

Dosages of prednisone in the 10- to 15-milligram range usually do not completely shut down the adrenal gland but make it sluggish; on this dosage, the adrenal cortex may make only 2 to 5 milligrams instead of 7.5 milligrams a day. Taking more than 15 milligrams of prednisone for a month usually shuts off the hypothalamic-pituitary-adrenal axis and no steroid is made by the body. Figure 9 (in Chapter 17) illustrates these concepts.

## Why Are There So Many Different Kinds of Steroids?

In addition to the numerous natural steroid derivatives, specially engineered synthetic steroids have become available. The most important preparations are listed in Table 13. Prednisone is the most commonly used oral preparation, followed by prednisolone and dexamethasone. Prednisolone is almost the same as prednisone, but some patients tolerate it better and it can be the steroid of choice for patients with liver disease. These agents are usually given once a day, but in the presence of acute inflammation they are metabolized (or used up) more quickly and must be given 2 to 4 times a day. Dexamethasone enters the central

**Table 13.** *Some Oral, Injectable, and Intravenous Steroid Preparations*

| Hormone Preparation | Equivalent in Milligrams | Comment |
|---|---|---|
| Cortisone acetate (Cortone) | 25 | Very short-acting; must be given every 6 hours |
| Hydrocortisone (Hydrocortone; Solu-Cortef) | 20 | Short-acting; works within minutes |
| Deflazacort | 6 | May be safer for bones and stomach lining; not yet available in the United States |
| Prednisone (Orasone; Deltra) | 5 | Takes 2 to 4 hours to work and lasts 18 hours; most popular preparation |
| Prednisolone (Hydeltra) | 4 | Almost the same as prednisone; preferred if liver disease is present |
| Methylprednisolone (Medrol) | 4 | Similar to prednisone and prednisolone; used intravenously as Solumedrol |
| Triamcinolone (Aristocort; Kenalog) | 4 | Popular as an injection and as a dermatologic preparation for rashes |
| Dexamethasone (Decadron) | 0.75 | Lasts 24 to 36 hours; good penetration in the central nervous system |
| Betamethasone (Celestone) | 0.6 | Popular as an injection and for rashes; matures fetal lungs in pregnant patients with lupus at risk for delivering prematurely |

nervous system better than prednisone or prednisolone and is the steroid of choice for neurosurgeons treating postoperative swelling. It has a longer duration of action: dexamethasone is prescribed no more often than once a day and sometimes just several times a week. A newer steroid preparation, deflazacort, is available in Europe and Mexico (as Calcort) and may have fewer side effects than the compounds listed above.

Oral steroids can be taken daily or on an alternate-day schedule. Taking twice the dose every other day has the theoretical advantage of decreasing the risk of infection and should only be considered once a patient's clinical condition is stabilized.

*Local steroids* are used for skin involvement of lupus. They are of two types, fluorinated and nonfluorinated, and they are available in several forms: creams, gels, solutions, ointments, and occlusive dressings. Fluorinated steroids are very potent; if used too often, they cause skin thinning and atrophy. They should never be applied to the face continuously for more than 2 weeks at a time. Nonfluorinated steroids are very safe but much weaker, can be purchased without a prescription (e.g., Cortaid), and are used for facial rashes. The most effective topical salves are ointments. Though oily and gloppy, they are 80 percent absorbed. Creams are the best tolerated. They dry the skin but are only 20 percent absorbed and are therefore less effective than ointments. A dermatologist can also deliver steroids into the skin by injection, or into lesions. This highly effective technique means using thin needles and injecting discoid lesions and areas of hair loss with steroid solutions.

*Pulse steroids* are administered in high doses intravenously. Examples include methylprednisolone (Solumedrol) and dexamethasone (Decadron). These are given in dosages equivalent to 1000 milligrams of prednisone per day to achieve high anti-inflammatory levels in critically ill patients. The effect of pulse steroids often lasts for weeks. Patients who cannot tolerate oral steroids may occasionally be given monthly pulse steroids. Hospitalized patients who cannot take any medications by mouth (e.g., after surgery) are given intravenous steroids in lower doses that simulate a corresponding oral dosage.

Steroids can also be given *intraarticularly* or *intramuscularly*—in other words, into a joint or muscle. Steroids injected into joints consist of preparations that have more of a water or oil base. Water-based preparations do not stay in the joint and quickly enter the bloodstream. These include hydrocortisone, dexamethasone, and methylprednisolone (e.g., Hydeltra-TBA and Depo-Medrol). Thicker preparations stay in joints for weeks (betamethasone or triamcinolone; Celestone, Kenalog, Aristocort), and one triamcinolone preparation, Aristospan, never really enters the bloodstream and stays in a joint for months. The choice of steroid depends on the patient's clinical picture: If the problem is limited to one joint, Aristospan is advised; but if a patient complains especially of knee pain but also hurts all over, Kenalog is used. Aristospan is not widely prescribed because of its expense—it costs five times more than some other brands. I am frequently asked how often steroids can be injected into joints, tendons, or ligaments. Studies have suggested that injecting a joint with a steroid more often than every 3 months destroys cartilage and is not advisable. Too many injections into tendons or ligaments can weaken or rupture these structures.

## What Types of Steroid Regimens Are Used to Treat Lupus?

Active, organ-threatening lupus should always be treated with steroids. Involvement of lupus in the heart, lung, kidneys, liver, or blood (autoimmune hemolytic anemia or thrombocytopenia) is managed with *high-dose steroids*—one milligram per kilogram (1 kilogram = 2.2 pounds) per day of prednisone for at least 6 weeks. This usually averages out to 40 to 80 milligrams a day. Doses above 40 milligrams a day are considered high. Acute central nervous system vasculitis is treated with high-dose steroids for shorter periods of time, often with additional pulse therapy (see the previous section).

Severe flareups of non-organ-threatening disease warrant *moderate doses,* or 20 to 40 milligrams of prednisone a day. This circumstance arises when a patient complains of severe chest or heart pains that turn out to be pleurisy or pericarditis; manifests active skin disease; or has a severe flare of joint inflammation. These moderate doses are maintained for a few days to several weeks.

Chronic, mild, non-organ-threatening disease responds to daily doses of 2 to 20 milligrams of prednisone. This is considered *low-dose therapy*. Most patients with non-organ-threatening disease on low- or moderate-dose steroids have other agents added and ultimately are switched over to NSAIDs, antimalarials, or methotrexate, all of which are safer than using long-term steroids unless the doses are very low.

## What Are Surgical and Stress Doses of Steroids?

The adrenal gland helps compensate the body for stress by releasing extra adrenalin (epinephrine and norepinephrine) and steroid. Surgery, for example, stresses the body, and we all need extra adrenalin and steroids to handle it. When lupus patients undergo surgery or face any stressful procedure, doctors usually give two to three times the usual daily dose of steroids on the day of the event to ease stress; sometimes it takes a few days before the cortisone doses are back to their original levels.

## What Can Steroids Do That's Bad?

Steroids are a blessing and curse. Without them, many lupus patients would die; but with them, serious complications can arise. Certain reactions to steroids occur immediately, while other reactions occur in patients who have had higher doses of the drugs or have taken lower doses for longer periods of time. One immediate reaction to steroids is a sense of energy. However, they can also cause palpitations, agitation, and rapid heart rates, and they can make it hard to sleep. Many of my patients tell me that they clean house at 3 A.M. after starting moderate-dose steroids. Heartburn is an early side effect in patients beginning steroids.

As time passes, numerous additional side effects occur that can best be categorized by organ systems. The *skin* becomes thin and wrinkles easily. Bruises appear, hair loss in the male pattern of baldness from the scalp is noted (in the temples and the upper back of the head), while facial hair increases, along with facial acne. Wound healing is impaired and sores take longer to disappear. Stretch marks are noted throughout the skin, especially on the abdomen. *Musculoskeletal* problems include muscle weakness and loss of calcium in bones (which can lead to osteoporosis). Clots of fat released by steroids can settle in the blood supply of bones, leading to a condition known as avascular necrosis or "dead bone." The skin and muscles become very sensitive to the touch and fibromyalgia may ensue.

Since steroids are hormones, the *endocrine/metabolic system* is affected. Among the complications caused by steroids are glucose intolerance, which eventually becomes diabetes; menstrual irregularity; and central obesity, where one's body weight is redistributed toward the belly and buttocks and away from the extremities. Steroids can stunt the growth of children taking these drugs, and—as discussed earlier—they can suppress the hypothalamic-pituitary-adrenal axis. The production of fats, especially cholesterol and triglycerides, is increased. Cataracts form and glaucoma develops in the *eyes*.

The *central nervous system* is stimulated, which produces agitation, confusion, difficulty in concentrating, and may even result in psychotic behavior. Patients may become moody and irritable. Some of the *cardiovascular-renal* effects lead to salt and water retention with bloating and puffiness, especially in the face and ankles. Potassium blood levels decrease, often necessitating oral replacement. Salt retention causes blood pressures to rise. The *gastrointestinal system* is involved, as continued steroid use causes the heartburn to lead to ulcers. Steroids thin intestinal walls and increase the risk of perforation, especially of diverticula (pouchings) in the colon. The pancreas can also become inflamed by corticosteroids. Finally, the risk of *infection* is much greater, since steroids prevent antibodies from killing bacteria, viruses, fungi, and parasites.

Not everybody develops all of these complications and many individuals who are steroid-dependent develop none of them. It is impossible to predict which complication will occur in whom, except that those who have taken these drugs in higher doses are more likely to note some of these side effects.

## How Does One Get off Steroids?

Unless a person has been on steroids for less than a month, the answer is slowly. Acute lupus flareups can be treated with short courses of steroids, given as a Medrol Dosepak or Aristocort Dosepak, where high doses are dropped to zero within a week. Or steroids can be injected into the buttocks with Celestone, Aristocort, Depo-Medrol, or Kenalog, where the drug is out of the system within several weeks. These short-term treatment courses are rarely if ever associated with serious problems.

The ideal way for a doctor to taper steroids if a patient has been on the drug for at least a month is to decrease the dose by 10 percent a week until adrenal replacement levels are reached (7.5 milligrams of prednisone daily). A *steroid withdrawal syndrome* simulating fibromyalgia frequently mimics disease flareups and must be differentiated from active lupus. This syndrome is self-limited and disappears within 2 to 3 weeks of keeping the steroid doses at the same level. When adrenal replacement levels are reached, the doctor may wish to taper more slowly (e.g., 10 percent reductions every few weeks or reduced doses of 5 percent every week or 10 days). On occasion, the doctor may order a *Cortrosyn stimulation test,* which evaluates the sluggishness of the adrenal glands and offers guidance as to how fast or slowly the medication can be tapered. Many of the side effects listed above, especially those of weight gain and bloating, can persist for several months after the drugs have been discontinued.

## How Do Doctors Minimize the Side Effects of Steroids?

Since doctors know what can happen when they start patients on cortisone, is there anything they can do to minimize the reactions that have just been discussed? The answer is yes, to a certain degree. For example, I sometimes prescribe antacids or ulcer medicines (especially $H_2$ blockers such as Tagamet, Zantac, or Pepcid) along with steroids. Patients are urged to keep to a low-sodium, low-fat, and low-carbohydrate diet and to limit their calorie intake. Sometimes, diuretics are prescribed to deal with bloating and fluid retention. Steroid-dependent patients should take in at least 1 gram of calcium daily to protect their bones. Mild sedatives allow patients to get some rest at night, and families are counseled to be supportive rather than angry at the appearance or behavior of their loved ones. I also urge my patients to keep active, which minimizes muscle atrophy and osteoporosis, and counsel them to stay away from friends or colleagues who have colds or other infections. No environment is ideal for a patient taking steroids, but it can be tailored to some extent to make life more comfortable.

## IMMUNOSUPPRESSIVE THERAPIES

"Chemotherapy" can be a scary-sounding word. Some rheumatologists instead use the terms "immunosuppressive," "cytotoxic," or "steroid-sparing" therapy. Essentially, these therapies are used when serious organ-threatening disease is present and steroids alone are not sufficient to deal with the problem. They are also employed when a patient cannot tolerate high doses of steroids, since chemotherapies reduce steroid requirements. Chemotherapies are not as scary as they sound and are very useful agents in the treatment of SLE.

In 1947, a patient with severe discoid lupus had *nitrogen mustard* ointment applied to the skin and the rashes healed. Shortly thereafter—and in the days before dialysis—this agent was given to patients with severe kidney disease and proved to be lifesaving. However, nitrogen mustard is messy and inconvenient; the doctor has to wear a Darth Vader–like space suit to apply the ointment, while intravenous mustard must be given in the hospital and patients become quite ill from it. Efforts were made to improve upon this useful compound, and this led to the introduction of *cyclophosphamide* in the early 1960s. Also known as Cytoxan, it is chemically similar to nitrogen mustard but much better tolerated. Until the early 1980s, Cytoxan was given as a daily pill. Even though it proved to be effective in managing severe lupus and was steroid-sparing, several problems arose. First of all, patients had to drink gallons of water daily to prevent a condition known as hemorrhagic cystitis. This was manifested by bloody urine and sometimes evolved into bladder cancer. Also, most patients lost their hair and had to wear wigs. Further, many young women stopped menstruating and became sterile. Cytoxan also upset the stomach and lowered blood counts, which made patients more susceptible to infection. Finally, Cytoxan use for several years was associated with a 5 to 10 percent risk of developing a malignancy in 10 to 20 years. Cytoxan and nitrogen mustard are alkylating agents; their effects simulate total-body irradiation. In the early 1980s, a kinder, gentler way of administering Cytoxan was found. This involved once-a-month intravenous administration. When Cytoxan was used in a cyclical fashion, the incidence of hemorrhagic cystitis dropped to less than 5 percent; patients did not lose their hair if they wore an ice cap or head tourniquet while getting the drug; newer medicines that minimized nausea became available (e.g., ondansetron, also known as Zofran); the malignancy risk decreased to less than 1 percent; and fewer patients became sterile. However, most female patients over the age of 30 who take at least nine monthly doses of cyclophosphamide in a 1-year period stop menstruating.

Recent studies have suggested that patients with organ-threatening disease who do not quickly respond to steroids benefit from the addition of intravenous, intermittent Cytoxan. For severe kidney disease (Chapter 19), doctors usually give Cytoxan once a month for 6 months and then every 2 to 3 months until 2 years have elapsed. Cytoxan does not always work; up to 30 percent of patients who take the drug also require the addition of a second agent or therapy such as Imuran or plasmapheresis.

*Chlorambucil* (*Leukeran*) is an oral alkylating agent commonly used in Europe and developing countries, but it is prescribed in the United States only for patients who are allergic to or cannot tolerate Cytoxan. It is much better tolerated than oral Cytoxan; however, it is less effective and more dangerous than intravenous Cytoxan because its long-term use makes it a potent carcinogen. When I prescribe chlorambucil, I make every effort to limit its use to less than 2 years.

In the early 1960s, an agent became available that prevented patients with kidney transplants from rejecting them. Not a chemotherapy, since it does not treat cancer, *azathioprine* (*Imuran*) has turned out to be a controversial addition to the rheumatologist's therapeutic arsenal. A slight modification of *6-mercaptopurine,* which is still rarely used today, Imuran blocks inflammation pathways and is FDA-approved for rheumatoid arthritis. Numerous studies examining the role of Imuran in SLE have been disappointing. It has modest effects on serious lupus but has the advantage of reducing steroid requirements. Some investigators are studying its use in combination with Cytoxan for kidney disease. I occasionally come across patients who refuse Cytoxan because it is more important for them not to compromise their chances of having children than it is to drastically reduce their risk of dialysis. Other patients choose Imuran because they are concerned about the carcinogenicity of alkylating agents.

Imuran ends up being used frequently as a compromise between potent chemotherapies and an ongoing requirement for high doses of oral steroids. Imuran is generally well tolerated; about 15 to 20 percent of patients have some stomach upset, which can be managed by dividing the doses or lowering them. Imuran does not alter fertility, and although it is not without risk, thousands of kidney transplant patients have had normal children while taking the drug (Chapter 30). Abnormal liver function tests and low blood counts can be found and should be monitored every month at the beginning of treatment and ultimately every few months. Imuran, like all immunosuppressive therapies, increases one's susceptibility to infection. Although Imuran can be carcinogenic after many years of use, administration of the drug for 2 to 4 years in patients with active SLE is not associated with increased cancer risk.

A popular rheumatoid arthritis drug, *methotrexate,* has been on the market since the 1940s. In high doses, it is a very effective agent in managing numerous types of cancer. The drug is popular because it works within weeks and is taken only once a week. Acting via its inhibition of folic acid, which is necessary for cell growth, low doses of methotrexate decrease joint inflammation and have been used in SLE. However, several problems have been encountered. First of all, methotrexate can make patients with lupus more sensitive to ultraviolet light. Also, it appears to have little effect on organ-threatening disease. Additionally,

**Table 14.** *Immunosuppressive Therapies Available to Treat SLE*

| |
|---|
| Alkylating agents |
| Cyclophosphamide (Cytoxan)—The most widely used agent. It can be effective, powerful, and toxic. |
| Nitrogen mustard—Almost the same as cyclophosphamide but messier to give and has to be administered in a hospital. |
| Chlorambucil (Leukeran)—Available only orally. Very well tolerated but probably more carcinogenic than cyclophosphamide or nitrogen mustard. |
| Antimetabolites |
| Azathioprine (Imuran)—A good steroid-sparing agent. Much weaker than alkylating agents. |
| Methotrexate (Rheumatrex)—A good steroid-sparing agent for lupus arthritis if sun sensitivity is minimal. Must be used carefully with kidney impairment; there is no evidence that it helps organ-threatening disease. |
| 6-Mercaptopurine—Almost the same as azathioprine; rarely used because it is more toxic to the liver. |

the drug is handled by the kidneys. Therefore, if renal disease is present, the doses have to be drastically reduced in order to prevent a precipitous drop in blood counts. Anyone taking methotrexate must refrain from drinking alcohol and must have blood and liver function tests made at regular intervals. Nausea, mouth sores, and headaches are the most common side effects of methotrexate. I restrict its use to lupus patients who have active joint inflammation without significant skin or organ-threatening disease.

Table 14 lists the major immunosuppressive therapies available to treat SLE.

## Summing Up

Antimalarials are prescribed for non-organ-threatening disease and to relieve skin, joint, pleural, and fatigue symptoms. They take many months to work and probably decrease the risk of lupus spreading to critical organs. I give them to all newly diagnosed lupus patients without organ-threatening disease for at least 2 to 4 years. Corticosteroids are used in high doses for critical organ disease; otherwise, function in that organ might be lost. They can be prescribed for mild to moderate disease, but antimalarials, NSAIDs, and methotrexate should be started at the same time, allowing steroids to be discontinued as soon as possible. Chemotherapies such as cyclophosphamide or azathioprine are used for patients with severe, active disease. They work with corticosteroids in decreasing disease activity and also allow steroid doses to be tapered. Corticosteroids and immunosuppressives are very toxic, and careful monitoring is necessary.

# 28

# *Other Options: Treatments Occasionally Used to Manage Lupus*

In addition to the disease-modifying drugs discussed in the last chapter, many other agents have a place in managing different aspects of lupus. These drugs or therapies "fall into the cracks," or have limited uses under special circumstances. Most of them have a "bridging" function: they are used for short periods of time until one of the disease-modifying agents discussed in the previous chapter becomes effective. This section offers a brief overview of these bridge therapies.

## NON–CONTRACEPTIVE HORMONES

The use of estrogen-containing hormones for contraceptive purposes has been the focus of many discussions of lupus. Far less attention has been given to other hormones that may be useful in SLE.

Since 90 percent of lupus patients are women, rheumatologists logically thought that male hormones might help them. *Testosterone* was given to female lupus patients without success as early as 1948, and numerous unsuccessful findings have been published since. This might stand to reason, since males with lupus often fare worse than females. However, evidence that estrogen decreases the immune response and male hormones increase it suggests that certain male hormone products might be useful. A male-type hormone known as *dehydroepiandrosterone* (*DHEA*) is a promising agent that is being studied in the treatment of mild to moderate lupus; it may be particularly useful in patients with cognitive dysfunction.

Even though there has been only one report in the world's literature of a lupus flare from *estrogens* given to manage menopausal symptoms (e.g., Premarin, Estrace, estrogen patches), many physicians are needlessly reluctant to prescribe these replacement hormones. This is partly because, unlike birth control pills (discussed in Chapter 30), replacement hormones are not associated with clinically relevant blood clotting abnormalities, probably because they have only 20 percent as much estrogen as birth control pills. These agents are generally tolerated very well by my patients and also have a place in treating *osteoporosis*. Osteoporosis is the loss of calcium in bones and is more common in lupus, due to chronic inflammation and corticosteroid therapy. Severe bone loss leads to painful bony compression fractures (Chapters 13 and 29).

*Danazol* (*Danocrine*), marketed for endometriosis, is an antiestrogen which, for unclear reasons, can improve autoimmune hemolytic anemia and low platelet counts associated with SLE. I use it for patients with low platelet counts who have no lupus disease activity elsewhere in the body. *Bromocriptine* (*Parlodel*) inhibits the secretion of breast milk by blocking the actions of prolactin, the hormone that stimulates the production of breast milk. Considerable evidence has accumulated indicating that many lupus patients have elevated prolactin levels. This, in turn, is associated with an immune dysfunction. Unfortunately, patients of mine who have taken bromocriptine have had no improvement in their clinical lupus.

## RETINOIDS

Retinoids are vitamin A derivatives that block cell growth and act as anti-inflammatories. Retinoids were not used in the past in part because they, with the exception of beta carotene, tend to increase sun sensitivity. But recent reports indicate that retinoids are very effective in managing subacute cutaneous lupus and refractory discoid lesions if they are used with sun blocks, and this has changed our thinking. Several of them, including *13-cis-retinoic acid* (*Accutane*), *etretinate* (*Tegison*), *and acitretin* (*Etretin—not available in the United States*), are potent agents usually used to treat psoriasis or acne, and they also have anti-lupus effects. All these agents can raise lipid levels (especially triglycerides) and are very drying. None of these drugs should be prescribed for women who might become pregnant, since they induce birth defects. Their effects last only as long as they are taken; they do not have the disease-modifying properties of the antimalarials. As with antileprosy drugs, retinoids are used concurrently with antimalarials until the rashes improve to the degree that they can be discontinued.

*Beta carotene* is a nonprescription retinoid health supplement that acts as a mild sun block but has little if any effect upon lupus.

## ANTILEPROSY DRUGS

Leprosy—a bacterial infection primarily involving the skin, joints, and peripheral nervous system—provokes a mild autoimmune reaction. Cutaneous leprosy and discoid lupus can be difficult to tell apart.

Antileprosy drugs have actions similar to those of antimalarial agents and are useful for resistant skin and joint problems. However, they have no place in treating serious systemic activity. *Dapsone* is the most commonly used drug in this class. It has been helpful in managing hard-to-treat skin and joint lupus, especially if a particular rash called bullous lupus is present. Bullous lupus resembles chickenpox except that its distribution is different and the blisters tend to be larger. If it is left untreated, serious systemic complications can develop, especially dehydration. Patients taking dapsone (who have an initial negative screening blood test for a abnormality known as G6PD) should have liver tests and blood counts monitored every few weeks. A G6PD deficiency is a contraindication to the use of dapsone since it can cause hemolytic anemia. *Clofazimine* (*Lamprene*) is a recent addition to the antileprosy armamentarium in the United States. Particularly useful for refractory discoid skin lesions, it has the advantage of being well tolerated and is taken only once a day. Unfortunately, Lamprene can stain the skin much as Atabrine does, although the pigmentation also disappears upon its discontinuation.

A third limited-use alternative, *thalidomide,* was introduced in Europe as a mild sedative in the late 1950s. A worldwide scandal exploded in the early 1960s when the agent was implicated in causing birth defects and numerous children were born with limbs missing in whole or in part. Patients who might become pregnant must not even consider trying this agent. However, the drug has remained available for leprosy patients who had no other alternatives. And in the mid-1970s, dermatologists in Mexico and Germany documented its benefits for antimalarial-resistant cutaneous lupus. Thalidomide has sedating effects and should be taken at night. Up to 10 percent of patients develop a peripheral neuropathy, which can be rather annoying, as it produces continuous numbness and tingling. Further, even though thalidomide clears most lupus lesions, its benefits usually wear off as soon as the drug is stopped. Most rheumatologists use antileprosy drugs along with antimalarials and stop the former as soon as the patient improves to the point that antimalarials alone can control the problem.

## DRUGS THAT INTERACT WITH CYTOKINES AND GROWTH FACTORS

Cytokines and growth factors are hormone-like substances that promote cell growth and modulate inflammation. Agents that interact with these chemicals can influence SLE. See Chapter 5.

*Cyclosporin A* (Sandimmune) is an expensive drug (costing $12,000 to $15,000 a year) that prevents organ transplant rejections in part by blocking the cytokine interleukin-2. Inflammatory rheumatoid-like arthritis responds to this drug, which, however, is rarely needed for this. Cyclosporin A was avoided in lupus for a long time because it raises blood pressure, causes unwanted hair growth, and elevates lipid and creatinine (kidney function) blood levels in tests. But, most lupus patients who take the drug after a kidney transplant have tolerated it surprisingly well with few adverse reactions. Some recent evidence suggests that class V (membranous) lupus nephritis responds particularly well to this agent. A newer, similar agent known as tacrolimus (Prograf, FK506) has recently become available.

*Alpha interferon* has been used locally to treat cutaneous lupus and is effective in managing lupus complicated by a protein known as a cryoglobulin. On the other hand, when given for chronic active hepatitis or hairy cell leukemia, this cytokine aggravates joint aching and inflammation in patients with preexisting lupus and worsens cognitive dysfunction. The cytokine *interleukin-2* is approved for advanced malignancies and can flare lupus as well as induce fibromyalgia.

Growth factors stimulate the production of red blood cells (*erythropoietin, or EPO*), white blood cells (*G-CSF*), or macrophages (*GM-CSF*) by the bone marrow and are occasionally used in lupus patients. Because the growth factor EPO is very well tolerated in SLE, it has been used to manage severe anemia, especially among dialysis patients. The other growth factors—G-CSF and GM-CSF—are usually employed when blood counts are very low (as a complication of chemotherapy) and the patient is a risk for life-threatening infections. These growth factors, however, can occasionally cause flareups of musculoskeletal lupus and produce allergic-type reactions.

## GOLD

This unlikely metal was first shown to have antirheumatic properties when Jacques Forestier tried it on tuberculosis patients in France in the 1920s. Patients who also happened to have rheumatoid arthritis got better. Gold was first used for lupus in the 1940s, when many patients who were diagnosed with rheumatoid arthritis really had SLE. Many of these individuals with lupus had allergic reactions and rheumatologists shied away from using gold for SLE for many years. Lupus is still listed as a contraindication by the manufacturer for injectable gold but not for oral gold. As different gold injection preparations were introduced along with gold pills, investigators took a new look at the drug in the 1980s. Indeed, gold is helpful in managing arthritis and cutaneous lupus and should be considered when other agents fail to benefit the patient. Oral gold is usually quite safe except for a mild diarrhea seen in half of the patients; gold shots and pills require monitoring of urine sediment to detect protein and blood counts. Patients may also develop allergic-type rashes and mouth sores.

## SPECIALIZED TREATMENTS FOR KIDNEY DISEASE

*Dialysis* involves using a machine to remove wastes from the blood—wastes that are normally cleared by the kidney. Patients with lupus and kidney failure make up about 5 percent of the SLE population and need dialysis in order to stay alive. This can be performed with *hemodialysis,* by which the patient's blood is cleansed three times a week for several hours, or *peritoneal dialysis,* which is often performed continuously at home, with a catheter connected in the abdominal area. Although moderately unpleasant and time-consuming, both procedures are usually well tolerated, but hemodialysis may be better at helping decrease lupus disease activity and allowing patients to discontinue steroids. The ideal management for end-stage renal disease is a *kidney transplant*. One-third of lupus patients on dialysis are able to undergo a transplant successfully. The best results are seen when the kidney donor is a sibling under the age of 60 who has normal blood pressure and matches the patient's blood type. Cadaver transplants are a second choice. Lupus only rarely returns to the transplanted kidney, and more than 85 percent of all transplants prove to be successful after 5 years of follow up.

Several preparations help kidney disease by decreasing urine protein leakage, improving blood flow to the kidney, and lowering blood pressure. These agents include *angiotensin converting enzyme* (*''ACE''*) *inhibitors* such as captopril. Other therapies, including prostaglandin $E_1$ infusions, may also turn out to be useful.

## IMMUNE STIMULANTS

Serious lupus is treated with immune suppressants. However, over the years, researchers have studied the effects of immune stimulants—which do the opposite—in SLE with disappointing results. For example, *Levamisole* is an agent available for managing colon cancer, but several large-scale trials with lupus patients during the 1970s failed to show any benefit for SLE. Similarly, *thymus gland* extracts and derivatives (e.g., thymosin) have not been useful.

## GAMMA GLOBULIN

Gamma globulin is a protein that circulates in the blood, boosts immunity, and helps fight and prevent infection. It has been used for years as an intramuscular injection with modest results. Once the technical problems were overcome in the late 1970s, much larger amounts of the immune globulin could be administered intravenously (IV). Intravenous gamma globulin has several mechanisms of action, among the most important of which is its ability to prevent the spleen from destroying platelets. It is very expensive (up to $5000 per treatment course) compared to gamma globulin shots ($20 per injection), but the former can be lifesaving.

However, IV gamma globulin is indicated only for short-term use in autoimmune thrombocytopenia (low platelets) until other agents can become effective. It may be used monthly for the occasional lupus patient who has recurrent infections and low gamma globulin levels (most lupus patients have elevated gamma globulin levels). The majority of published studies to date suggest that IV gamma globulin has no place in the management of fibromyalgia or chronic fatigue.

## LASERS

There isn't much to say about lasers yet. Laser therapy for discoid lesions or inflamed joints is still being researched but holds promise for the future.

## THERAPEUTIC BLOOD OR LYMPH GLAND FILTERING

As you recall, lymphocytes are white blood cells that help control the immune response. Since steroids and chemotherapy work partly by destroying lymphocytes, it seemed logical that removing lymphocytes from the body with a filtering machine might be effective and reduce inflammation with fewer side effects. In the late 1960s, investigators found that draining lymphocytes from the thoracic duct (a lymph gland collection site near the left collarbone) was feasible. At the modest cost of $250,000 per patient, *thoracic duct drainage,* as it was called, drained billions of lymphocytes from patients and decreased disease activity for about 4 months. This led researchers to attempt the removal of lymphocytes directly from the blood, a process called *lymphapheresis.* Performed on a blood filtering device, *apheresis* (meaning "to remove by force" in Greek) was difficult to perform in SLE patients. This is because they had antilymphocyte antibodies which, along with steroids and chemotherapy, decreased lymphocyte counts to such a low level that the machine could not remove enough cells to make a difference. Investigators at Harvard University and Stanford University then attempted to irradiate lymphocytes in the lymph glands. Known as *total lymphoid irradiation* (*TLI*), it resulted in improvements that lasted for several years. However, 80 percent of the patients developed shingles; they also lost an average of 20 pounds and 10 teeth. Furthermore, the use of TLI prevented further useful chemotherapy from being employed when the disease ultimately flared. (Prior radiation mixed with alkylating agents such as cyclophosphamide or chlorambucil greatly increases a patient's risk for developing cancer.)

Attention was then shifted to depleting plasma on a cell separator device that resembles a dairy creamer. Removing plasma from whole blood and returning red blood cells, white blood cells, and platelets to the patient is called *plasmapheresis.* Since many proteins and antibodies promote inflammation, plasmapheresis was tried in a variety of critical lupus complications. This expensive procedure ($50,000 to $100,000 for a several-weeks therapeutic course) is beneficial and lifesaving for some lupus patients with thrombotic thrombocytopenic purpura (TTP), hyperviscosity syndrome, and cryoglobulinemia. However, attempts to use plasmapheresis for serious organ-threatening complications of lupus such as nephritis or central nervous system disease have generally been disappointing, although occasional results have been dramatically successful. Researchers are currently manipulating concurrent medications, replacement products, and plasmapheresis schedules in order to improve its efficacy.

In spite of all the scary sounding procedures mentioned in this section, the technical aspects of plasma or lymphocyte removal are generally well tolerated, with only a 3 percent rate of serious complications, which is very good considering that doctors are dealing with seriously ill patients. Apheresis produces some temporary light-headedness, dizziness, and cramping; moreover, vascular access—the ability to find a vein for intravenous infusions—can be a problem. Apheresis is indicated for TTP, hyperviscosity syndrome, and cryoglobulinemia and may be helpful in patients with life- or organ-threatening complications that do not respond to steroids or chemotherapy.

## Summing Up

Even though NSAIDs, antimalarials, corticosteroids, and chemotherapy account for the overwhelming majority of all lupus therapies, other treatments are occasionally useful. Antileprosy drugs help resistant skin lesions and joint inflammation; retinoids are also useful for refractory skin rashes. These preparations are only helpful temporarily and act as a bridge to antimalarials. Cytokines and growth factors must be used carefully, since adverse reactions are common. The only hormones that have a role in SLE are danazol for low platelet counts and certain anemias and postmenopausal estrogen replacement therapy. Immune stimulants are disappointing. Gamma globulin helps fight immune deficiencies, which promote infection, and can raise low platelet counts. Machines can act as artificial kidneys or be used to remove antibodies, cells, or proteins from the body. The role of maintenance dialysis is firmly established, but many uses of apheresis are still controversial.

# 29
# *Fighting Infections, Allergies, and Osteoporosis*

In most recent surveys, infection has now taken second place as one of the most common causes of death among lupus patients, after cardiovascular complications. Why are lupus patients so susceptible to infection and what can be done about it? Should at-risk patients receive vaccinations, or would that make the disease worse? And what about allergies and allergy shots? Moreover, one of the major problems facing women after they go through the change of life is osteoporosis, and women with lupus are especially vulnerable to this disease. This chapter outlines a practical approach for dealing with infections and vaccines while promoting awareness of the issues involved in infection, allergy, and lupus. Finally, the management of osteoporosis is reviewed as it applies to lupus.

## WHAT IS AN INFECTION?

Patients with systemic lupus have altered abilities to fight common infections and are susceptible to attack by uncommon organisms that rarely affect healthy people. Infections of this kind are called *opportunistic*. Opportunistic infections are usually seen only in cancer patients receiving chemotherapy, individuals suffering from AIDS, and those who have altered immune systems. Autoimmune diseases fall into the last category, particularly when immunosuppressives or steroids are being used.

Microbes can attack cells and, in turn, kill cells and damage tissue. The body produces antibodies to these microbial antigens, killing foreign organisms as part of an anti-inflammatory and occasionally an allergic response. Many different kinds of organisms can infect the body. Depending on size, structure, life-cycle, and ability to produce certain chemicals, they are called bacteria, viruses, parasites, fungi, or protozoans. Common infections, from streptococcal or staphylococcal bacteria, affect healthy people all the time.

## ARE LUPUS PATIENTS MORE VULNERABLE TO INFECTION?

Since 80 percent of patients with SLE take corticosteroids during the course of their disease and 10 to 30 percent undergo chemotherapy, it stands to reason that lupus patients are vulnerable to infectious processes. The NSAIDs and antimalarials do not increase susceptibility to infection. Even if you are not receiving steroids or chemotherapy, you are still susceptible to unusual infections. Lupus patients attacked by common microbes frequently require treatment for longer periods of time and often project a more severe clinical picture. Their susceptibility is changed in several ways. First, the microbial killing function of neutrophils, a type of white blood cell, is altered, and, second, certain complement components critical in the killing process are inhibited. Third, there is evidence to indicate that many lupus patients have circulating blood factors which inhibit a component of complement (C5) from being attracted to an inflamed cell. Finally, cytokine dysfunction in SLE (Chapter 5) decreases the body's ability to kill foreign organisms.

Additionally, corticosteroids block cells from destroying other cells and thus from diminishing inflammation. Chemotherapies prevent the reproduction of cells with immunologic memories, which can signal the body to kill microbes and decrease the numbers of cells themselves.

## WHAT KIND OF INFECTIONS MIGHT A LUPUS PATIENT GET?

The principal bacterial infections seen in lupus affect the respiratory tract (*Streptococcus, Staphlococcus*) and urinary tract (*Escherichia coli*). As a lupus patient, you could also be at special risk for developing infections in your joints (septic arthritis), tuberculosis, or salmonellosis. In addition to cold viruses, you might be especially susceptible to herpes zoster, or shingles. Epstein-Barr virus, hepatitis viruses, and cytomegalovirus may be slightly more prevalent in SLE. The most common fungal process in lupus patients is *Candida,* or yeast. This cheesy-white infiltrate is seen in the throat and esophagus (causing sore throat and difficulty swallowing) and in the vagina. Rarely, unusual fungal infections such as those caused by *Cryptococcus, Coccidioides immitis,* and *Nocardia* are observed. In the United States, the principal parasitic pathogen seen in lupus patients is called *Strongyloides,* and the protozoan to be aware of is *Pneumocystis carinii,* which is frequently found in AIDS patients.

## HOW ARE INFECTIONS DIAGNOSED?

Active lupus mimics infection, and infectious processes not only mimic lupus but can flare it. Infections can be difficult to diagnose. The principal manifestations of infections include fever, sweats, and shaking chills. If you have a high fever for more than a few days, you should be thoroughly evaluated, especially if you are also taking NSAIDs or steroids, which lower the body temperature. I try to take a thorough medical history, noting symptoms and inquiring about recent travel (especially abroad), exposure to illness, purchase of a new pet, and occupation-related illnesses. I often order a throat, urine, blood, or stool culture in addition to a complete blood count or chest x-ray. Other causes of fevers such as active lupus, cancer, allergy, or a drug reaction must be considered. Some physicians find a C-reactive protein (CRP) blood test helpful in differentiating active lupus from infection, but this is controversial. Anybody with a suspected life-threatening infection from an unknown source may have to be observed and have cultures taken in a hospital where gallium scanning, bone marrow biopsy, lymph node biopsy, or lung bronchoscopy can be performed to make a critical diagnosis as safely and quickly as possible and where intravenous antibiotics are readily available.

## HOW ARE INFECTIONS TREATED?

Many rheumatologists have noted that patients with SLE need higher doses of antibiotics for longer periods of time than healthy people do. Most infections can be treated on an outpatient basis, but serious bacterial processes and most opportunistic infections necessitate a period of in-hospital intravenous antimicrobial therapy. Antimicrobial agents fight numerous types of organisms. Antibiotics attack bacteria and prevent them from reproducing. Antiviral, antiparasitic, antiprotozoan, and antifungal drugs are also available. Although steroids may delay the response to antibiotics, the two can be taken together, since lupus frequently flares up when the patient becomes infected.

## CAN INFECTIONS BE PREVENTED?

Some prevention is possible and just requires common sense. For example, a lupus patient's exposure to people with colds or other infections should be minimized. Antibiotics are sometimes given to such patients as a preventive measure before, during, or after certain surgical and dental procedures. A patient's dentist should know that he or she has lupus. Dental procedures in those with Libman-Sacks endocarditis are managed with at least 1 gram of ampicillin or amoxicillin 2 hours before and 4 to 6 hours afterwards. If a patient is allergic to penicillin, there are substitutes available. I occasionally encounter lupus patients who develop repeated colds or infections. For these individuals and others at high risk (e.g., patients who work at day-care centers or are nursery or kindergarten teachers), I boost immunity with a gamma globulin shot every 3 to 4 months. While there is no proof that this is effective, I have found that it decreases the intensity, duration, and number of minor infections. During flu season, exposure to a person with influenza usually warrants a 48-hour course of amantadine (Symmetrel, 100 milligrams twice a day) or rimantadine (Flumadine, 100 milligrams a day for 2 days) to prevent the development of symptoms. The key to prevention is to be aware of the conditions in your environment and make sure your doctors and dentist know you have lupus so they can take necessary steps to protect you.

## WHAT ABOUT VACCINES?

Occasional reports have appeared of healthy persons developing lupus after receiving a routine vaccination. The concept of vaccinating patients with small amounts of a provoking substance, with the goal of having them make antibodies to an infectious agent, dates back to Edward Jenner's experiments with smallpox in the 1700s. Nearly all individuals with SLE have been vaccinated against a variety of diseases with little difficulty. Some vaccines use live organisms (e.g., virus) and others do not. Vaccines against measles, mumps, and polio, among others, use live viruses. Even though there is a theoretical risk in exposing a lupus patient to a live virus or to a family member who has received a live-virus vaccine, there has never been a case report of the patient contracting the disease. Passive immunization with nonspecific antibodies such as gamma globulin poses no problems in patients with SLE. On the other hand, I have frequently been asked whether patients can develop lupus from a vaccine. While this is theoretically possible, it is a very rare occurrence and probably happens in no more than one out of several thousand genetically predisposed persons.

Difficulties have been encountered with some types of immunizations in lupus patients. Some investigators have observed that certain patients who received tetanus or flu vaccines, for example, also made antibodies to DNA or other lupus autoantigens. It seems that flu vaccines do not work as well if the patient has SLE; antibodies achieve only half the desired levels for half as long. Additionally, up to 20 percent of those with SLE may feel sick or achy for a few days, which is double the incidence in the general population. These patients should consult their physicians before they receive any vaccine. Some rheumatologists give their lupus patients flu shots; others chose to prevent influenza with amantadine or rimantadine (antiviral antibiotics) when their patients are exposed. As a general practice, I do not give routine flu shots to patients with lupus. In potentially serious circumstances, however, rheumatologists rarely hesitate to give necessary vaccinations.

## ARE LUPUS PATIENTS MORE LIKELY TO HAVE ALLERGIES?

Generalized allergies or increased sensitivities to environmental chemicals or drugs are observed in 10 percent of the general population and in 20 to 25 percent of those with SLE. Patients with lupus are more liable to have drug, insect, and skin allergies as well as asthma. What does this mean for the patient? First of all, allergic lupus patients have no difficulty using the antihistamines, inhalers, nasal sprays, Sudafed-containing decongestants, or steroids that many otherwise healthy allergy-prone patients take from time to time. On the other hand, sulfa antibiotics can make them more sun-sensitive and cause more allergic reactions; they should be avoided if possible. Also, a minority of lupus patients who receive *allergy shots* (*immunotherapy*) make more autoantibodies and experience disease flareups. For this reason, in 1989, the World Health Organization recommended that patients with autoimmune diseases should *not* receive allergy shots. Nonspecific allergy shots might cause them to make more anti-DNA and other lupus-related antibodies in addition to making antibodies against the offending allergen.

## WHAT IS OSTEOPOROSIS?

With the passage of time, all of us are likely to have some thinning of our bones. Manifested by a loss of calcium in bone mineral, this can lead to fractures, bone pain, and shorter stature. The consequence can be an inability to live independently. Lupus patients are especially susceptible to this demineralization process, or osteoporosis.

We have two types of bone: cortical (as in the hips) and trabecular (as in the vertebrae of the spine). With age, men and women become osteoporotic, or lose calcium in their cortical bones. Additionally, the onset of menopause selectively demineralizes trabecular bone. Women are more susceptible to osteoporosis in general. Persons who are Caucasian, have thin builds, abuse tobacco or alcohol—as well as those with a genetic predisposition—are also at increased risk. The use of corticosteroids accelerates this process, which is why osteoporosis is so common in SLE. Further complicating this picture is the general hesitancy of doctors to prescribe hormones to lupus patients and the belief that certain chemotherapies can bring on premature menopause.

## HOW IS OSTEOPOROSIS MANAGED IN SYSTEMIC LUPUS?

Despite all the gloomy risk factors detailed above, an intelligent woman with lupus who works with her health-care team can frequently prevent the serious complications induced by osteoporosis. I usually order a *bone mineralization study* in women at risk to help me decide what management is optimal. These studies go under several names, such as QDRs, QCTs, DEXAs, or dual-photon absorptiometry. They are inexpensive, painless, take 15 minutes to perform, and provide a lot of information. I usually order them at 1- to 2-year intervals in high-risk patients.

Women with early SLE should initiate preventive measures such as taking 1 to

1.5 grams of calcium by mouth daily in divided doses. Along with a well-balanced diet, preparations such as Tums EX, Os-Cal, and Posture are easily obtainable without a prescription. A regular exercise program, as reviewed in Chapter 24, also decreases demineralization rates. The role of postmenopausal replacement estrogen therapy is reviewed in Chapter 28. These agents are well tolerated in lupus. If significant osteoporosis is present and the bone mineral study suggests that there is a high risk of spontaneous fracture, I frequently intervene with vitamin D, fluoride, or a bisphosphonate such as Didronel. The recommended doses and schedules of these drugs are in continuous flux; therefore a doctor should be consulted. Finally, the use of a calcium hormone called calcitonin may be advised. This agent is currently available only as an injection (it is similar to administering insulin; one has to be taught to self-inject it). In the near future this highly effective agent will be marketed as a nasal spray or perhaps as an oral preparation. The most important thing one can do regarding osteoporosis is to be aware of the risk of this condition and try to prevent it.

## Summing Up

Lupus is complicated by an increased susceptibility to infection, and this is associated with a greater risk of disease-related complications. A rheumatologist or family physician should be consulted (at least by telephone) before a patient with SLE takes an antibiotic or receives a vaccine. Prudent preventive measures help prevent problems later. When there is a question of serious, life-threatening infection, careful evaluation (and perhaps hospitalization) is essential. Patients with lupus are more vulnerable to allergies, drugs, insects, and chemicals. Extreme caution should be exercised before taking allergy shots or sulfa antibiotics. Patients who have risk factors for osteoporosis should arrange to have themselves tested and, if necessary, treated, since this condition is more common with lupus.

# 30

# *Can a Woman with Lupus Have a Baby?*

Most women with SLE want to have children. This is certainly understandable, since 90 percent of patients with lupus are female and 90 percent develop the disease during their reproductive years. Unfortunately, many of my colleagues advise lupus patients not to become pregnant on the basis of incorrect or outdated information. In addition, doctors are notorious for lacking a special sensitivity that is often needed when pregnancy is ill advised. The good news is that the overwhelming majority of lupus patients can have normal babies. On the other hand, there are circumstances when a pregnancy presents increased risks. This chapter attempts to confront and clarify misconceptions and notions often held by both patients and doctors.

## DOES LUPUS ALTER THE ABILITY TO CONCEIVE?

The ability to procreate, or fertility, is usually normal in SLE. Discoid lupus and drug-induced lupus per se are not associated with fertility problems. But several specific circumstances relevant to SLE may affect fertility. These include disease activity, dialysis, and drugs.

Patients with very active disease frequently have irregular periods or none at all. This is the body's reaction to stress; menstrual regularity is restored when the disease is under adequate control. Thirty percent of patients with SLE have significant kidney disease, and 10 to 20 percent of them evolve end-stage renal disease, necessitating dialysis over a 10-year observation period. Dialysis is associated with scanty or no menstrual cycles. A successful kidney transplant obviates the need for dialysis and may restore regular periods. Finally, certain chemotherapy drugs interfere with ovulation (Chapter 27). Cyclophosphamide causes premature menopause in 25 to 50 percent of the women in their twenties who receive it for at least a year and in 80 percent of the women who take it in their later thirties. Azathioprine, cyclosporine A, methotrexate, and nitrogen mustard are not associated with loss of periods in women with SLE. Investigational studies under way would enable women who need to take cyclophosphamide to have their eggs removed and stored for later use.

What about male fertility? Sperm counts (but not libido) decrease in men who take chemotherapies. Although sterility is uncommon, I urge males with SLE who require chemotherapy to bank (store) their sperm before starting treatment.

## THE MOTHER: WHAT WILL GO ON INSIDE OF HER?

### The Overall View

In 1991, our group conducted a survey and chart review of 307 women with SLE who had 634 pregnancies and were seen in our office between 1980 and 1990. Eighty patients (26 percent) never conceived; this is three times the national average. Of the 634 pregnancies, 439 (69 percent) resulted in live births, 106 (17 percent) in therapeutic abortions, and 95 (15 percent) in spontaneous abortions. These patients fell into low-risk, high-risk, and moderate-risk categories.

### Low-Risk Mothers

Those who fall into this group have nothing to worry about, since their risk of a problem pregnancy is the same as that in the general population (5 to 10 percent). These women include those with discoid lupus, drug-induced lupus, and women with SLE who have mild disease and are in remission, off all medication, and lack the Ro (SSA) antibody and anticardiolipin antibody.

### High-Risk Mothers

A small group of women face an extremely high risk of fetal demise and maternal organ failure if they become pregnant. Overall, up to 3 percent of high-risk pregnant women with SLE die. This group includes patients with active lupus myocarditis, active lupus nephritis with an elevated serum creatinine, severe and uncontrollable high blood pressure, and those who need to receive chemotherapy during their pregnancy. Myocarditis is usually aggravated during pregnancy and leads to heart failure. Many lupus patients with active nephritis and an elevated serum creatinine will require dialysis, and that hikes the risk of fetal demise to over 80 percent. Preexisting hypertension is aggravated during most pregnancies, and, if not well controlled, can lead to strokes. Finally, most chemotherapies other than cyclosporine A or azathioprine have an unacceptably high risk of causing fetal anomalies, malformations, and maternal infections. Despite these odds, I continuously come across brave women who wish to exercise their biological prerogative and don't care that the odds are greatly stacked against them. About 20 percent of the mothers make it through, have a normal baby, and come out unscathed. Most mothers with these risk factors miscarry; the remaining 20 percent develop serious complications. Lupus patients with serious, organ-threatening disease who wish to have children should be encouraged to consider adoption.

## Moms Who Should Proceed with Caution

Patients who are not at especially high risk but for whom a pregnancy might well present problems fall into the category that most of my young female patients fit into. This section provides a clear road map to follow.

### *What Will Happen to the Lupus?*

Patients whose lupus is mild or moderately active at the time of conception have a 40 percent chance of having no change in their disease, a 40 percent chance of flaring, and a 20 percent chance of improving. The fetus makes cortisone, and by the second trimester, mild disease may improve as the mother receives this extra dose of steroids. However, various chemicals released in pregnancy can also promote inflammation. Most flares are mild and easy to manage. Serious flares rarely occur in this group, but mild cutaneous or musculoskeletal postpartum flares are common, especially between the second and eighth weeks after delivery. The withdrawal of fetal steroids from the body may have something to do with this. Although few rheumatologists do this, I routinely give SLE patients an injection of 60 milligrams of triamcinolone (Kenalog) intramuscularly 7 to 10 days postpartum to block the flare.

### *What Lupus Medications Can Patients Take During a Pregnancy?*

*Regular-dose aspirin or other NSAIDs* are not advisable during a pregnancy since they may induce bleeding, which leads to miscarriage, prolongs labor, and causes early closure of the opening between the pulmonary artery and descending aorta near the fetal heart. However, these dangers are not applicable to patients taking low-dose aspirin who have antiphospholipid antibodies. A patient who is pregnant and has inadvertently taken an NSAID on a short-term basis and has not miscarried has no cause for concern, since fetal malformations do not occur.

*Antimalarials* are not advisable in pregnancy, and the manufacturers list pregnancy as a contraindication. The administration of chloroquine is associated with a very small risk of causing blindness or deafness in the fetus. Several reports of congenital malformations with quinacrine have appeared. In my experience, however, hydroxychloroquine (Plaquenil) appears to be safe but has the theoretical risk of causing the same hearing and sight problems as chloroquine. I do not recommend a therapeutic abortion if Plaquenil was taken early in the pregnancy, since these complications have never been reported. I used to recommend that patients discontinue all antimalarials as soon as they try to conceive; this counsel now applies only to chloroquine or quinacrine. I advise my Plaquenil patients to stop their medication as soon as they find out they are pregnant.

Moderate or low-dose *corticosteroids* (less than 40 milligrams of prednisone a day) appear to be relatively innocuous and free of significant problems in pregnancy. This is probably because they are natural hormones made by both mother and fetus. A steroid preparation called betamethasone (sometimes injected into joints as Celestone) crosses the placenta particularly well and is used by obstetricians to improve the maturity of fetal lungs in mothers who show signs of delivering prematurely.

The use during pregnancy of any of the agents discussed in this paragraph is not advised by the manufacturers. *Azathioprine* rarely, if ever, causes fetal abnormalities and has been taken during pregnancies by patients who have had kidney transplants. Occasional reports of fetal immune deficiencies for the first months have appeared. Another transplant antirejection agent, *cyclosporine A,* appears to be relatively safe. *Cyclophosphamide, methotrexate, nitrogen mustard,* and *chlorambucil* are all capable of causing fetal malformations and should be avoided unless the life of the mother is at stake and she wishes to continue the pregnancy.

*Plasmapheresis* and *intravenous gamma globulin* are safe to use in pregnancy.

### *How Can Complications Related to the Antiphospholipid Syndrome Be Prevented?*

The antiphospholipid syndrome (Chapter 21) is associated with an increased risk of fetal death and miscarriage. One-third of those with SLE have antiphospholipid antibodies (especially anticardiolipin), and in these patients the risk of spontaneous abortion ranges from 20 to 50 percent. *Antiphospholipid antibodies* cross the placenta and promote clots in it, which result in fetal death. I usually screen all newly pregnant patients for antiphospholipid antibodies if they have not been tested previously. Occasionally, doctors come across a patient who is anticardiolipin-negative but becomes positive only during the pregnancy.

There is no agreed upon way to manage pregnant women with antiphospholipid antibodies. If the antibody is present, I prescribe one baby aspirin a day (81 milligrams) during the pregnancy until the 35th week, when it is stopped to allow the baby's heart channel to close. There is a much greater risk that IgG antibodies to cardiolipin will induce abortion than that IgM or IgA antibodies will do so, and the IgG antibodies can be tested for easily. If the mother miscarries in spite of baby aspirin therapy, the next time she conceives, I initiate therapy with subcutaneous heparin, a blood thinner injected twice daily under the skin, like insulin. Oral warfarin (Coumadin), which is usually used when patients with these antibodies have systemic blood clots while taking baby aspirin, is not advised in pregnancy. Many respected rheumatologists use moderate doses of prednisone (20 to 40 milligrams daily) along with baby aspirin, and still others use plasmapheresis in patients (wash their blood to filter out the antibody) weekly or give intravenous immunoglobulin. My success rate with baby aspirin—and if needed, subcutaneous heparin—is over 50 percent, and I have only occasionally needed to resort to steroid or blood-filtering therapies.

### *What Should Patients Who Carry the Ro (SSA) Antibody Do?*

Between 20 and 30 percent of those with SLE carry the Ro (SSA) antibody. Many of these patients also carry the La (SSB) antibody; the presence of this antibody by itself is very rare. Anti-Ro and anti-La can cross the placenta and induce two syndromes: neonatal lupus and congenital heart block. (Both are discussed in detail in Chapter 23.) These antibodies present no risk to the mother. The chance of developing either of these syndromes is very small, and cutaneous neonatal lupus is a mild, self-limited process. Even though congenital heart block in the infant is found in less than 5 percent of pregnancies of Ro-positive mothers, pregnant women should be screened for it with a fetal echocardiogram (ultrasound of the heart) during the second trimester.

## What about Patients with Active Kidney Disease?

Renal disease present at the beginning of pregnancy is associated with a 50 percent flare rate, a 25 percent incidence of preeclampsia (pregnancy-related hypertension), and a 25 percent risk of kidney failure requiring dialysis during the pregnancy. Patients who have normal renal function (creatinines of less than 1.5 milligrams per deciliter) should be closely monitored, with doctor visits approximately every 2 weeks. Rigid blood pressure control, salt restriction, and increased doses of steroids if renal function worsens or serum complements fall are desirable. Patients with elevated serum creatinines are at an even greater risk and must follow these precautions. If the nephritis was well controlled prior to conception, only 10 percent of patients have serious flareups during pregnancy.

## How Should Pregnant Patients Feel and What Should They Do?

Pregnant patients with lupus behave and feel like most healthy pregnant women. There is no reason why they cannot work or exercise if they wish to and are able. No special dietary considerations apply. There is a chance that if they have mild to moderate disease, their lupus symptoms could worsen during the first trimester. The fetus starts making cortisone during the second trimester, and the mother will probably start feeling better at that time. The child's father should be included on the health-care team so that he can help and support the mother when she needs rest or doesn't feel up to doing things. Flareups can be treated with acetominophen (Tylenol) if fever or musculoskeletal aching is involved and with steroids if they are more serious.

### What Laboratory Tests Should Be Obtained during a Lupus Pregnancy?

When a patient tells me she is pregnant, I ask her (and the father) to come into my office so I can explain the situation and answer many of the questions raised in this chapter. In addition to examining the patient, I also perform baseline laboratory studies. These include a complete blood count, platelet count, blood chemistry panel, urinalysis, anticardiolipin antibody, complement studies, anti-DNA, electrocardiogram, and anti-Ro (SSA) and anti-La (SSB) antibodies. The Westergren sedimentation rate is falsely elevated in all normal pregnancies and is not reliable. I try to see my pregnant lupus patients once each trimester and more often if necessary. Blood pressures and weight are monitored at each visit. Pregnancy is associated with a physiologic anemia that occurs as red blood cells are diluted out with the increased volume of body fluids. Therefore, the development of anemia must be significant in order to be a concern. At follow-up visits, I check a complete blood count, dipstick the urine (which screens for protein and sugar), and obtain C3 and C4 complement levels. A fall in complement values is an excellent indicator of disease flares in pregnancy. Additional testing is done only if the patient's complaints or medical history warrant it. The patient's doctor should make every effort to communicate regularly with her obstetrician so that they can work together as a team.

### What Can the Patient Who Is Breast-Feeding Take?

Up to 20 percent of the NSAIDs taken by a mother who breast-feeds reaches the infant and may cause a bleeding tendency or acidosis; therefore, these agents are not advised. Even though only 1 percent of antimalarial drugs are excreted in breast milk, toxic levels in the baby can be reached very quickly. It would also be dangerous to expose any infant to chemotherapy unnecessarily. The only antilupus drug that is safe to use with breast-feeding is prednisone. Studies have shown that doses of up to 30 milligrams daily taken by the mother have no untoward effects on the baby. The advantage of breast-feeding an infant applies only for the first 3 months of life. Breast-feeding after that time limits the physician's ability to intervene in managing lupus activity and is not advisable unless the patient is in a complete remission.

## THE FETUS

### What Are the Chances That the Baby Will Have Lupus?

The risk that any child born to a mother with SLE will develop the disease is small. The "lupus gene" or sets of genes predisposing to lupus has what doctors call "a low penetrance." In other words, fewer than 10 percent of patients who carry a lupus gene will ever develop the disease.

If anti-Ro (SSA) and anti-La (SSB) are absent, the risk of being born with lupus is one in several thousand. These rare occurrences are associated with the anti-RNP antibody (see Chapters 6 and 11 for a review). Cutaneous neonatal lupus is seen in less than 5 percent of patients with anti-Ro or anti-La and is a self-limited, benign process that disappears within weeks to months. Congenital heart block is discussed in Chapter 22.

The chance that an offspring of a lupus patient will develop the disease in childhood and adult life is 10 percent for females and 2 percent for males. However, up to 50 percent will carry autoantibodies in their blood (especially ANA) and up to 25 percent will develop an autoimmune disease (including lupus) in their lifetimes. The most common non-lupus autoimmune process is thyroiditis, which is usually mild and benign, followed by rheumatoid arthritis.

## What Are the Chances of a Successful Pregnancy?

A recent literature review concluded that among nonterminated pregnancies of mothers with SLE in the United States between 1980 and 1991, the frequency of miscarriage was 17.5 percent; stillbirth, 6.9 percent; prematurity, 29 percent; and neonatal death, 8 percent. The last category included babies who died within 30 days of birth. All told, two-thirds or 67 percent of all lupus pregnancies produce a successful birth. This is considerably less than the national rate of 85 percent. Most of the miscarriages can be traced to the antiphospholipid antibodies which, if identified, can lead to successful treatment during the next pregnancy. High blood pressure, systemic lupus activity, and gestational diabetes from steroid therapy also contribute to prenatal deaths. Prematurity, defined as birth before the 37th week of gestation, is quite common in SLE cases and is usually associated with an active maternal disease. There are, however, other factors for premature births, as evidenced by many of my healthy mothers who deliver early.

## Is Abortion Safe and When Should It Be Considered?

Contrary to what some medical texts stated in the 1960s, abortion poses no special risks for lupus patients. However, the decision to terminate a pregnancy should not be taken lightly, and moral, religious, ethical, financial, and social considerations must be discussed with the patient and her family. Speaking personally, the only times I advise an abortion is when the life or critical well-being of the mother is at stake. This usually occurs when the only way to reverse serious organ-threatening disease, reduce hypertension, or prevent the mother from going on dialysis is to terminate the pregnancy.

## Family Planning and Birth Control

I've noted that, barring unusual circumstances, lupus patients are normally fertile. Women with SLE should ideally plan their pregnancies during a period of disease remission or relative inactivity that has been sustained for several months. Barrier methods of contraception such as diaphragms or condoms are generally effective. Though less efficacious, spermicidal creams, sponges, or jellies are also safe. Intrauterine devices (IUDs) should be used with caution, since they are associated with an increased risk of pelvic infection. Since lupus patients have difficulty fighting infection, this could be problematic. Progesterone-containing contraceptives such as Micronor do not have any specific lupus-related complications or concerns and are safe in SLE.

## Can Lupus Patients Take Birth Control Pills?

When estrogen-containing birth control pills first became available in the early 1960s, there were numerous reports of young women either developing lupus or having their preexisting lupus flare while taking these agents. By the early 1970s, these reports completely disappeared. Nevertheless, many physicians trained during this period who have not kept up with the rheumatology literature advise their lupus patients against taking oral contraceptives. What was responsible for the initial reports? First of all, birth control preparations in the 1960s contained much more estrogen than they do today. Also, estrogen-containing contraceptives during that era had tartrazine preservatives, which are no longer used. As discussed in Chapter 8, aromatic amines such as tartrazines are known to induce lupus.

Even though hundreds of my lupus patients have taken birth control preparations without difficulty, there are certain individuals who should not take them. *Estrogen-containing contraceptives are not recommended for patients with SLE who have antiphospholipid antibodies, high blood pressure, migraine headaches, a history of abnormal blood clotting, or very high lipid (cholesterol or triglyceride) levels.* These clinical subsets are associated with an increased risk of developing blood clots or having strokes. Beyond these groups, there is no evidence that contemporary oral contraceptives induce lupus in a genetically susceptible patient. *Progesterone-containing contraceptives do not present any specific risks in SLE.* Also, several surveys have shown that ANAs do not develop as a result of taking birth control pills. Some published papers have suggested that estrogen-containing contraceptives induce disease flares in a minority of patients, but other studies failed to show any relationship. It appears that mild, reversible flares can occur in a small number of lupus patients taking birth control pills.

### The Delivery and Postpartum Period

Mothers and their obstetricians should aim for a vaginal delivery. The incidence of cesarean sections is the same as in the general population except for unusually high-risk medical situations (e.g., uncontrollable blood pressure). Once the placenta is delivered, steroid levels drop. After 2 to 8 weeks, the body goes into a steroid withdrawal. If that extra fetal steroid boost was suppressing the lupus, this is the time when lupus flares. In order to prevent this, as noted above, I frequently see my lupus patients 10 to 14 days after delivery and give them 60 milligrams of Kenalog intramuscularly. This modestly dosed timed-release steroid prevents steroid withdrawal symptoms from occurring and decreases the risks of a postpartum flare.

## Summing Up

Lupus patients are normally fertile unless they have very active systemic disease or have received chemotherapy. Lupus pregnancies are successful 67 percent of the time; 13 percent fail due to antiphospholipid antibodies and up to 30 percent of all deliveries are premature. If frequent competent clinical evaluations, blood testing, and good obstetric care are available, most lupus patients with mild to moderately active disease do quite well. Oral contraceptives should not necessarily be ruled out in family planning. Steroids can be safely used for exacerbations of lupus during pregnancy or for postpartum flares in those who choose to breast-feed.

# 31
# *What's the Prognosis?*

When I first diagnose a patient with systemic lupus, she frequently inquires about the bottom line. She wants to know outcomes. Will I live? Can I have children? Is this crippling or deforming? Many lupus patients have excellent outcomes, live normal lives, and generally feel well. Others have serious impairments. I have to tell my patients that the prognosis, or bottom line, varies from individual to individual, and is dependent on many factors.

## WE'VE COME A LONG WAY!

I still come across newly diagnosed patients who are convinced that lupus is a fatal disease and that they only have 6 months to live. This nonsense is based on perceptions by some older physicians, who have had limited recent exposure to lupus patients. It also stems from the many outdated encyclopedias and reference books that line our family bookshelves. The first lupus survival study was published in 1939; it stated that half these patients were dead within 2 years. This 50 percent survival figure has constantly been revised. In 1955, half the patients were dead within 4 years, and by 1969, within 10 years. *In fact, at the present time, more than 90 percent of all lupus patients live more than 10 years,* although—if organ-threatening disease is present—only 60 percent survive 15 years.

How can we account for these improvements? First of all, the discovery of the LE prep in 1948 and ANA in 1957 allowed milder cases to be diagnosed, which increased the number of people with the disease. This increased the pool of SLE patients, and those with mild cases would naturally have longer lifespans. Also, the widespread availability of steroids by 1950, chemotherapy by 1955, and dialysis by 1970 extended life spans. Though less dramatic, additional factors have also greatly improved survival over the last 20 years: doctors have newer antibiotics to treat a variety of infections, better agents to attack high blood pressure, and a broader experience in managing disease complications.

## OUTCOME AND TYPE OF LUPUS

As you might expect, certain types of lupus have better outcomes than others. Cutaneous (discoid) lupus and drug-induced lupus are associated with a normal life expectancy. Patients with non-organ-threatening SLE can expect the same outcome as that of a person without lupus; more than 95 percent survive at least 10 years.

The "prognostic variables," as lupus specialists like to call them, depend on several major considerations. *Methods of health-care delivery* have an impact on determining outcome. For example, a Veterans Administration lupus study would largely restrict itself to males. Several Kaiser-Permanente HMO lupus surveys limited themselves to analyzing trends among middle class, insurable, working families. Hospital-based surveys focus on a sicker population of lupus patients. Publicly funded patients (e.g., Medicaid, county hospital patients) clearly have worse outcomes. Bluntly stated, if patients can read or write, afford food and medications, have transportation to clinics and access to rheumatologists who know their cases, and have been educated about the disease, their prognosis improves. *Epidemiologic factors* include race, geographic environment, and age. For example, more Asians (especially Chinese) develop SLE, but African Americans have a more severe process. Native Americans (American Indians) are also particularly prone to develop lupus. Lupus is rare on the African continent but observed in 1 of 300 African American women. Childhood SLE is more organ-threatening and severe; lupus is much milder in patients over the age of 60. Environmental considerations reveal the influence of climate, occupation, diet, and exercise; none of these has been adequately surveyed in lupus.

*Genetic factors* influence outcome. Males make up only 10 percent of all lupus patients but may have more severe disease. Also, a family history of autoimmune disease increases awareness of lupus, which can lead to earlier diagnosis and intervention.

*Clinical and laboratory variables*—findings from blood and other testing—suggest that organ-threatening disease has a worse outcome. High blood pressure, low platelet counts, anemia, elevated cholesterols, and renal involvement—especially with nephrotic syndrome—require more attentive and aggressive care. The presence of certain autoantibodies such as Ro (SSA), anticardiolipin antibody, and RNP can be seen as distinguishing different types of lupus with distinct treatments and outcomes.

Finally, *treatment variables* influence the course of the disease. Many rheumatologists rarely prescribe chemotherapies for kidney disease and restrict their patients to corticosteroids. When these investigators publish survival studies, they must be interpreted differently than those of centers that intervene liberally with azathioprine or cyclophosphamide. There is not right or wrong treatment, only different outlooks and viewpoints. More subtle "adjunctive" measures—such as whether or not doctors prescribe birth control pills, immunizations, antimalarials, special diets, and so forth—can also affect quality of life and survival.

## DOES LUPUS EVER GO AWAY ON ITS OWN?

Every few months, I come across a newly diagnosed lupus patient who listens patiently to my 30-minute speech on the disease, followed by specific therapeutic recommendations, and says, ''Doctor, I appreciate everything you are telling me, but I am going to take vitamins and herbs, eat a well-balanced diet, exercise, and see what happens.'' What happens to these patients? Can their disease go away on its own? Spontaneous remissions were reported as early as 1954. Non-organ-threatening disease has a 2 to 10 percent disappearance rate without medication. These are long-shot odds and I do not recommend taking the chance, since non-organ-threatening disease also has a 20 percent chance of becoming organ-threatening within 5 years, and I believe that antimalarial therapies further decrease this risk. The spontaneous disappearance of lupus in the heart, lung, kidney, liver, or hematologic systems is so rare that when one well-documented patient with severe lupus prayed to Father Junipero Serra (founder of the Spanish missions in California in the 1700s) and had her organ-threatening disease disappear, this evidence was submitted to the Vatican, where Serra was (and still is) being considered for sainthood.

In non-organ-threatening disease, lupus tends to burn itself out with time. Also, the onset of menopause is associated with milder disease. I occasionally allow reluctant patients with long-standing mild lupus to avoid medicine unless they are at risk for major complications.

## WHAT DO LUPUS PATIENTS DIE OF?

The natural course of SLE has been extensively studied and researchers have come to some interesting conclusions. The concept of a ''bimodal survival curve'' was first proposed in the 1970s and subsequently validated. This means that some lupus patients who die from the disease do so within the first 2 to 3 years of developing it. These individuals have active, aggressive lupus that responds poorly to therapy. After the third year, however, there's a lengthy hiatus of 10 to 15 years with few lupus-related deaths. But at 15 to 20 years, the effects of years of disease and medication seem to catch up with some patients, and a second mortality ''hump'' is observed. For example, young women who have active disease and are given moderate to high doses of steroids when they are 20 do well for a while, but eventually may experience complications from ongoing steroid therapy. This leads to diabetes, high blood pressure, elevated cholesterols, and obesity—which may result in heart attacks by the time they are 40, as in Bonnie's case (Chapter 14).

More than 90 percent of lupus patients with SLE die from one of five causes: complications of kidney disease, infections, central nervous system lupus, blood clots, or cardiovascular complications. For unknown reasons, fatal cancer is rare among lupus patients. Several trends have become evident since survival curves were first published in the 1950s. Improved methods of dialysis and the introduction of transplantation have substantially decreased kidney-related deaths. Superior methods of detecting and managing central nervous system lupus have also greatly decreased mortality from this complication. Unfortunately, infections are still a major cause of death, especially among patients receiving steroids or chemotherapy. The discovery of antiphospholipid antibodies in the 1980s and the use of blood thinning to prevent serious clots and strokes in patients at risk have not yet had an impact on survival in lupus, but improvements will be evident by the year 2000. Some types of lupus are still very difficult to treat, and insufficient progress has been made to improve survival. These subsets include those patients with pulmonary hypertension, mesenteric vasculitis, and TTP (thrombotic thrombocytopenic purpura). Bimodal survival curves are still relevant, but many fewer lupus patients are dying in the first 2 years.

## Summing Up

The outcome of SLE depends on who is treating the disease; which ethnic, racial, or geographically defined populations have the disease; and their socioeconomic status and therefore their access to subspecialty care; and the treatment philosophy of the health-care provider. In any case, more than 90 percent of all lupus patients in the United States live more than 10 years after being diagnosed. The survival of patients with organ-threatening disease is still an unsatisfactory 60 percent at 15 years. Patients with high blood pressure, low platelet counts, kidney disease, and severe anemia have a poorer outcome and should be managed aggressively. Mild lupus occasionally disappears spontaneously; serious lupus may ease up but does not go away without treatment. Deaths from lupus generally occur early on from active disease or later from continuously active inflammation or complications of therapy. Finally, in spite of all that has been said, I have found that patients who have a positive attitude and good coping mechanisms have a better prognosis.

# 32

# *Lupus in 2010: A Medical Odyssey*

What advances will take place in the next 15 years? Will we be able to cure lupus or prevent it? Is there anything to look forward to? Let's take a look at what the future holds—and it is indeed promising!

If developments proceed at the expected rate, my crystal ball suggests that by the year 2010, an integrated health-care system will be in place, allowing all lupus patients to receive optimal treatment regardless of socioeconomic status or medical insurability. A national data network should reveal exactly how many lupus patients there are as well as their gender and their racial, ethnic, and occupational background. The gene or combination of genes that predispose one to SLE and the environmental factors (viruses, chemicals, drugs, etc.) that turn these genes on will be known. It should be possible to identify individuals at risk for developing the disease and perhaps to vaccinate them so as to prevent autoimmune reactions. By 2010, we will know why 90 percent of patients with lupus are women, and we'll be able to manipulate hormones to decrease the disease's severity.

Immunologically, doctors and research scientists will probably have found all the subtypes of white blood cells there are and will have explored their functions and interactions. As part of their never-ending quest for knowledge, immunologists will describe additional autoantibodies and will better understand the ones we currently know about. Poorly understood clinical subsets of SLE, such as the lupus anticoagulant and central nervous system disease, will be firmly characterized and their pathogenic autoantibodies well delineated.

Existing therapies for lupus will be fine-tuned and improved upon. An ideal NSAID that treats mild inflammation without any adverse reactions, which may already be on the horizon, will be marketed. New-generation antimalarials and steroids that eliminate most of the side effects we associate with these agents should soon be available.

Our current chemotherapy approaches are very general: they suppress all types of white blood cells and do not substantially focus on any single "bad guy" subset. Increased use of combinations of chemotherapies that act at different levels of the inflammatory and immune process will be commonplace. The major advances in lupus therapy will emphasize cellular and antibody manipulations. Vaccines against particular antibody components are showing promise in animal models. These approaches encompass vaccinating patients to create antibodies against antigen-binding sites (idiotypes), T-cell receptors, IgM, T cells, and B cells. Special vaccines with specifically designed payloads can not only alter immunity as we know it but can also create new immunologic environments.

Albert Einstein said that God does not play dice with the universe. Some astounding developments now support this philosophical remark. The concept of apoptosis, or programmed cell death, has taken center stage in the last few years. Investigators in immunology have shown that a mouse model of lupus has an altered apoptosis gene which perpetuates certain aspects of inflammation and diminishes when the alteration is corrected. Is lupus related to some grand design?

Lupus is characterized by "dyslexic T cells," which leads to alterations in immune trafficking. These alterations will be manipulated with cytokine (e.g., interferons, interleukins) blockers, or antagonists, or cytokine receptor antagonists. Cell trafficking is facilitated by adhesion molecules, and new chemicals that block these adhesion molecules and decrease inflammation are now being developed. The effects of dyslexic T cells can also be reined in by inducing "tolerance." Chemicals that promote antigen-induced tolerance are being tested in mice with lupus.

Our concepts of cell immunology change every year. Some ethicists and philosophers in our discipline postulate that a higher authority created an immunotheology that can be manipulated only with rigid discipline if it is to be of any help to patients. We may hope that, by the year 2010, some lupus can be prevented, no one will die from it, and treatment will be both effective and safe.

# Glossary

**ACR** The American College of Rheumatology. A professional association of 4000 American rheumatologists of whom 2800 are board-certified. Criteria, or definitions for many rheumatic diseases, are called the **ACR criteria.** Formerly known as the ARA (American Rheumatism Association).

**Acute** Of short duration and coming on suddenly.

**Adrenal glands** Small organs, located above the kidney, that produce many hormones, including corticosteroids and epinephrine.

**Albumin** A protein that circulates in the blood and carries materials to cells.

**Albuminuria** A protein in urine.

**Analgesic** A drug that alleviates pain.

**Anemia** A condition resulting from low red blood cell counts.

**Antibodies** Special protein substances made by the body's white cells for defense against bacteria and other foreign substances.

**Anticentromere antibody** Antibodies to a part of the cell's nucleus; associated with a form of scleroderma called CREST (see its listing).

**Anticardiolipin antibody** An antiphospholipid antibody.

**Anti-double-stranded DNA (Anti-DNA)** Antibodies to DNA; seen in half of those with systemic lupus and implies serious disease.

**Anti-ENA** Old term for extractable nuclear antibodies, which largely consist of anti-Sm and anti-RNP antibodies.

**Antigen** A substance that stimulates antibody formation; in lupus, this can be a foreign substance or a product of the patient's own body.

**Anti-inflammatory** An agent that counteracts or suppresses inflammation.

**Antimalarials** Drugs originally used to treat malaria that are helpful for lupus.

**Antinuclear antibodies (ANA)** Proteins in the blood that react with the nuclei of cells. Seen in 96 percent of those with SLE, in 5 percent of healthy individuals, and in most patients with autoimmune diseases.

**Antiphospholipid antibody** Antibodies to a constituent of cell membranes seen in one-third of those with SLE. In the presence of a co-factor, these antibodies can alter clotting and lead to strokes, blood clots, miscarriages, and low platelet counts. Also detected as the lupus anticoagulant.

**Anti-RNP** Antibody to ribonucleoprotein. Seen in SLE and mixed connective tissue disease.

**Anti-Sm** Anti-Smith antibody; found only in lupus.

**Anti-SSA,** or the Ro antibody, is associated with Sjögren's syndrome, sun sensitivity, neonatal lupus, and congenital heart block.

**Anti-SSB,** or the La antibody, is almost always seen with anti-SSA.

**Apheresis** See Plasmapheresis.

**Apoptosis** Programmed cell death.

**Artery** A blood vessel that transports blood from the heart to the tissues.

**Arthralgia** Pain in a joint.

**Arthritis** Inflammation of a joint.

**Aspirin** An anti-inflammatory drug with pain-killing properties.

**Autoantibody** An antibody to one's own tissues or cells.

**Autoimmunity** Allergy to one's own tissues.

**Autoimmune hemolytic anemia** See hemolytic anemia.

**B lymphocyte or B cell** A white blood cell that makes antibodies.

**Biopsy** Removal of a bit of tissue for examination under the microscope.

**Bursa** A sac of synovial fluid between tendons, muscles, and bones that promotes easier movement.

**Butterfly rash** Reddish facial eruption over the bridge of the nose and cheeks, resembling a butterfly in flight.

**Capillaries** Small blood vessels connecting the arteries and veins.

**Cartilage** Tissue material covering bone. The nose, outer ears, and trachea consist primarily of cartilage.

**Chronic** Persisting over a long period of time.

**CNS** Central nervous system.

**Collagen** Structural protein found in bone, cartilage, and skin.

**Collagen vascular disease (also called connective tissue disease)** Antibody-mediated inflammatory process of the connective tissues, especially the joints, skin, and muscle.

**Connective tissue** The ''glue'' that holds muscles, skin, and joints together.

**Complement** A group of proteins that, when activated, promote and are consumed during inflammation.

**Complete blood count (CBC)** A blood test that measures the amount of red blood cells, white blood cells, and platelets in the body.

**Corticosteroid** Any natural anti-inflammatory hormone made by the adrenal cortex; also can be made synthetically.

**Cortisone** A synthetic corticosteroid.

**Creatinine** A blood test that measures kidney function.

**Creatinine clearance** A 24-hour urine collection that measures kidney function.

**CREST syndrome** A form of limited scleroderma characterized by C (calcium deposits under the skin), R (Raynaud's phenomenon), E (esophageal dysfunction), S (sclerodactyly or tight skin), and T (a rash called telangiectasia).

**Crossover syndrome** An autoimmune process that has features of more than one rheumatic disease (e.g., lupus and scleroderma).

**Cutaneous** Relating to the skin.

**Cytokine** A group of chemicals that signal cells to perform certain actions.

**Dermatologist** A physicisn specializing in skin diseases.

**Dermatomyositis** An autoimmune process directed against muscles associated with skin rashes.

**Discoid lupus** A thick, plaquelike rash seen in 20 percent of those with SLE. If the patient has the rash but not SLE, he or she is said to have **cutaneous (discoid) lupus erythematosus.**

**DNA** Deoxyribonucleic acid. The body's building blocks. A molecule responsible for the production of all the body's proteins.

**Enzyme** A protein that accelerates chemical reactions.

**Erythema** A reddish hue.

**Erythematous** Having a reddish hue.

**Estrogen** Female hormone produced by the ovaries.

**Exacerbations** Symptoms reappear; a flare.

**False-positive serologic test for syphilis** A blood test revealing an antibody that may be found in patients with syphilis and that gives false-positive results in 15 percent of patients with SLE. Associated with the lupus anticoagulant and antiphospholipid antibodies.

**Fibrositis, or fibromyalgia** A pain amplification syndrome characterized by fatigue, a sleep disorder, and tender points in the soft tissues; can be caused by steroids and mistaken for lupus, although 20 percent of those with lupus have fibrositis.

**Flare** Symptoms reappear; another word for exacerbation.

**FANA** Another term for ANA.

**Gene** Consisting of DNA, it is the basic unit of inherited information in our cells.

**Glomerulonephritis** Inflammation of the glomerulus of the kidney; seen in one-third of patients with lupus.

**Hematocrit** A measurement of red blood cell levels. Low levels produce anemia.

**Hemoglobin** Oxygen-carrying protein of red blood cells. Low levels produce anemia.

**Hemolytic anemia** Anemia caused by premature destruction of red blood cells due to antibodies to the red blood cell surface. Also called **autoimmune hemolytic anemia.**

**Hepatitis** Inflammation of the liver.

**HLA, or histocompatibility antigen** Molecules inside the macrophage that binds to an antigenic peptide. Controlled by genes on the sixth chromosome. They can amplify or perpetuate certain immune and inflammatory responses.

**Hormones** Chemical messengers—including thyroid, steroids, insulin, estrogen, progesterone, and testosterone—made by the body.

**Immune complex** An antibody and antigen together.

**Immunofluorescence** A means of detecting immune processes with a fluorescent stain and a special microscope.

**Immunity** The body's defense against foreign substances.

**Immunosuppressive** A medication such as cyclophosphamide or azathioprine, which treats lupus by suppressing the immune system.

**Inflammation** Swelling, heat, and redness resulting from the infiltration of white blood cells into tissues.

**Kidney biopsy** Removal of a bit of kidney tissue for microscopic analysis.

**La antibody** Also called anti-SSB, a Sjögren's antibody.

**LE cell** Specific cell found in blood specimens of most lupus patients.

**Ligament** A tether attaching bone to bone, giving them stability.

**Lupus anticoagulant** A means of detecting antiphospholipid antibodies from prolonged clotting times.

**Lupus vulgaris** Tuberculosis of the skin; not related to systemic or discoid lupus.

**Lymphocyte** Type of white blood cell that fights infection and mediates the immune response.

**Macrophage** A cell that kills foreign material and presents information to lymphocytes.

**MHC** Major histocompatibility complex; in humans, it is the same as HLA.

**Mixed connective tissue disease** Exists when a patient who carries the anti-RNP antibody has features of more than one autoimmune disease.

**Natural killer cell** A white blood cell that kills other cells.

**Nephritis** Inflammation of the kidney.

**Neutrophil** A granulated white blood cell involved in bacterial killing and acute inflammation.

**NSAID** Nonsteroidal anti-inflammatory drug, or agent that fights inflammation by blocking the actions of prostaglandin. Examples include aspirin, ibuprofen, and naproxen.

**Nucleus** The center of a cell that contains DNA.

**Orthopedic surgeon** A doctor who operates on musculoskeletal structures.

**Pathogenic** Causing disease, or abnormal reactions.

**Pathology** Abnormal cellular or anatomic features.

**Pericardial effusion** Fluid around the sac of the heart.

**Pericarditis** Inflammation of the pericardium.

**Pericardium** A sac lining the heart.

**Photosensitivity** Sensitivity to ultraviolet light.

**Plasma** The fluid portion of blood.

**Plasmapheresis** Filtration of blood plasma through a machine to remove proteins that may aggravate lupus.

**Platelet** A component of blood responsible for clotting.

**Pleura** A sac lining the lung.

**Pleural effusion** Fluid in the sac lining the lung.

**Pleuritis** Irritation or inflammation of the lining of the lung.

**Polyarteritis** A disease closely related to lupus featuring inflammation of medium and small-sized blood vessels.

**Polymyalgia rheumatica** An autoimmune disease of the joints and muscles seen in older patients with high sedimentation rates who have severe aching in their shoulders, upper arms, hips, and upper legs.

**Polymyositis** An autoimmune disease that targets muscles.

**Prednisone; prednisolone** Synthetic steroids.

**Protein** A collection of amino acids. Antibodies are proteins.

**Proteinuria** Excess protein levels in the urine (also called albuminuria).

**Pulse steroids** Very high doses of corticosteroids given intravenously over 1 to 3 days to critically ill patients.

**Raynaud's disease** Isolated Raynaud's phenomenon; not part of any other disease.

**Raynaud's phenomenon** Discoloration of the hands or feet (they turn blue, white, or red) especially with cold temperatures; a feature of an autoimmune disease.

**RBC** Red blood cell count.

**Remission** Quiet period free from symptoms, but not necessarily representing a cure.

**Rheumatic disease** Any of 150 disorders affecting the immune or musculoskeletal systems. About 30 of these are also autoimmune.

**Rheumatoid arthritis** Chronic disease of the joints marked by inflammatory changes in the joint-lining membranes, which may give positive results on tests of rheumatoid factor and ANA.

**Rheumatoid factor** Autoantibodies that react with IgG; seen in most patients with rheumatoid arthritis and 25 percent of those with SLE.

**Rheumatologist** An internal medicine specialist who has completed at least a 2-year fellowship studying rheumatic diseases (see above).

**Ro antibody** See anti-SSA.

**Scleroderma** An autoimmune disease featuring rheumatoid-type inflammation, tight skin, and vascular problems (e.g., Raynaud's).

**Sedimentation rate** Test that measures the precipitation of red cells in a column of blood; high rates usually indicate increased disease activity.

**Serum** Clear liquid portion of the blood after removal of clotting factors.

**Sjögren's syndrome** Dry eyes, dry mouth, and arthritis observed with most autoimmune disorders or by itself (primary Sjögren's).

**Steroids** Usually a shortened term for corticosteroids, which are anti-inflammatory hormones produced by the adrenal cortex or synthetically.

**STS** False-positive serologic test for syphilis.

**Synovial fluid** Joint fluid.

**Synovitis** Inflammation of the tissues lining a joint.

**Synovium** Tissue that lines the joint.

**Systemic** Pertaining to or affecting the body as a whole.

**T cell** A lymphocyte responsible for immunologic memory.

**Tendon** Structures that attach muscle to bone.

**Temporal arteritis** Inflammation of the temporal artery (located in the scalp) associated with high sedimentation rates, systemic symptoms, and sometimes loss of vision.

**Thrombocytopenia** Low platelet counts.

**Thymus** A gland in the neck area responsible for immunologic maturity.

**Titer** Amount of a substance, such as ANA.

**Tolerance** The failure to make antibodies to an antigen.

**Urinalysis** Analysis of urine.

**Urine, 24-hour collection** The collection of all urine passed in a 24-hour period; it is examined for protein and creatinine to determine how well the kidneys are functioning.

**UV light** Ultraviolet light. Its spectrum includes UVA (320 to 400 nanometers), UVB (290 to 320 nm), and UVC (200 to 290 nm) wavelengths.

**Uremia** Marked kidney insufficiency frequently necessitating dialysis.

**Vasculitis** Inflammation of the blood vessels.

**WBC** White blood cell count.

# Appendix
# Lupus Resource Materials

## WHAT ORGANIZATIONS FUND LUPUS RESEARCH, PROVIDE PATIENT SUPPORT, OR DISSEMINATE INFORMATION IN THE UNITED STATES?

American College of Rheumatology (ACR), 60 Executive Park South, Suite 150, Atlanta, GA 30329. Tel: 404-633-3777. The professional organization to which nearly all rheumatologists belong.

Arthritis Foundation, 1314 Spring St. NW, Atlanta, GA 30309. Tel: 404-872-7100 or 800-283-7800. Seventy-two American chapters provide research monies, publish literature, and provide patient support. The Arthritis Foundation spends about $2 million a year on lupus research.

LE Support Club, 8039 Nova Court, North Charleston, SC. Tel: 803-764-1769. Publishes a newsletter with many letters from patients, exchanging information.

Lupus Foundation of America (LFA), 4 Research Place, Suite 180, Rockville, MD 20850-3226. Tel: 301-670-9292 or 800-558-0121. Over 100 chapters and subschapters with 47,000 members. Publishes excellent pamphlets and provides patient support; local and national groups together provide slightly less than $1 million a year for lupus research. Dr. Carr's handbook is particularly good.

Lupus Network, Inc., 230 Ranch Drive, Bridgeport, CT 06606. Tel: 203-372-5795. Keeps an up-to-date on-line listing of all lupus resources and publishes a newsletter (*The Heliogram*) with critical reviews of these resources.

National Arthritis and Musculoskeletal and Skin Diseases Information Clearinghouse (AMS Information Clearinghouse), Box AMS, 9000 Rockville Pike, Bethesda, MD 20892. Tel: 301-495-4484. Resource service supported by the National Institutes of Health for rheumatic disease information including lupus.

National Institute of Arthritis and Musculoskeletal and Skin Diseases (NIAMS), Building 31, Room 4C05, 9000 Rockville Pike, Bethesda, MD 20892. Part of the National Institutes of Health, it funds several million dollars in lupus research each year at Bethesda and elsewhere, sponsors conferences, and publishes pamphlets. See the AMS Information Clearinghouse above.

The American Lupus Society (TALS), 260 Maple Court, Suite 123, Ventura, CA 93003. Tel: 805-339-0443 or 800-331-1802. Has about 30 chapters and 10,000 members, mostly in California and the Pacific Northwest. Provides patient support, pamphlets, and newsletters; supports about $100,000 in research annually. The brochures by Leslie Epstein, Graham Hughes, and Daniel Wallace are especially recommended.

## HOW CAN I FIND OUT ABOUT LUPUS SUPPORT OUTSIDE OF THE UNITED STATES?

The Lupus Foundation of America has an International Associates organization with 73 nations as members. Call the LFA for information. Some important organizations are listed below:

Lupus Canada, Box 322, Station B, Calgary, Alberta T2M 4L8, Canada. Tel: 800-661-1468, in Canada only.

British SLE Aid Group, ''Rookery Nook,'' 17 Monkhams Drive, Woodford Green, Essex, 1G8 0LG, England. Tel: 709-834-9365.

United Kingdom Lupus Group, 5 Grosvenor Crescent, London, SWIX 7ER, England. Tel: 01-235-0902.

Club de Lupus Centro Medico de Occidente, Pedro Buzeta 870-B, 44660 Guadalajara, Jalisco, Mexico. Tel: 41.43.05 or 16.01.32.

## HOW CAN I ORDER A GOOD VIDEO ON LUPUS?

A catalog is available from the AMS Clearinghouse, the Lupus Foundation of America, the Lupus Network, and some chapters of the Arthritis Foundation. See their listings above.

## WHAT ARE THE BEST BOOKS ON LUPUS WRITTEN BY NONPHYSICIANS?

Henrietta Aladjem's books are an inspiration. They include *Understanding Lupus* (Charles Scribner's Sons, 1985, 287 pp.), *A Decade of Lupus* (Lupus

Foundation of America, 1991, 178 pp.), *In Search of the Sun,* with Peter Schur (Charles Scribner's Sons, 1988, 264 pp.). Can be ordered from the LFA. See her foreword in this book.

Terry Nass's *Lupus Erythematosus: A Handbook for Nurses* (1985, 54 pp.) is very good for health-care workers. It can be ordered from LFA or TALS.

P. R. H. Phillips, *Coping with Lupus* (Avery Publishing, 1991, 256 pp.). Written by an eminent psychologist, this is the best book on the subject. Available from the LFA.

## ARE THERE OTHER LUPUS BOOKS WRITTEN BY DOCTORS FOR PATIENTS?

Three rheumatologists have recently written monographs for lupus patients:

Sheldon Blau's (with D. Schultz) *Living with Lupus* (Addison Wesley, 1993, 263 pp.) is an update of his 1977 effort, *Lupus: The Body Against Itself.*

Mark Horowitz's (with M. Brill) *Living with Lupus* has the same title as Blau's book (Plume, 1994, 201 pp.).

Robin J. Dibner, (with C. Colman) *The Lupus Handbook for Women* (Fireside, 1994, 176 pp.).

## WHAT ABOUT RHEUMATOLOGY OR LUPUS TEXTBOOKS?

Dr. Peter Schur and Dr. Marian Ropes wrote outstanding lupus books in 1983 and 1976 which have not been updated. The best current lupus books are:

D. J. Wallace and B. H. Hahn, *Dubois' Lupus Erythematosus,* 4th ed. (Lea & Febiger, 1993, 955 pp.). It can be ordered by dialing 800-638-0672.

Another excellent lupus textbook is R. G. Lahita, *Systemic Lupus Erythematosus,* 2nd ed. (Churchill Livingstone, 1992, 1022 pp.).

The best general rheumatology textbooks are:

W. N. Kelley et al., *Textbook of Rheumatology,* 4th ed. (Saunders, 1993, 1942 pp.).

D. J. McCarty and W. J. Koopman, *Arthritis and Allied Conditions: A Textbook of Rheumatology* (Lea & Febiger, 1993, 2100 pp.).

J. H. Klippel and P. A. Dieppe, *Rheumatology* (Mosby, 1994).

# Index